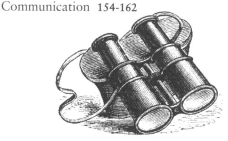

Introduction

Engineer Edwin Budding hit upon the idea of a lawn mower when he saw a machine used in the textile industry for shearing the nap from woollen cloth. He adapted the machine for cutting grass and in 1830 the lawn mower was invented and gardening made easier. In 1886, when frustrated at the number of plates her servants broke, the American housewife Josephine Cochrane devised the first mechanical dishwasher. Yet the ingenuity required to meet a need is nothing new, for even in 700 BC the Etruscans were making false teeth carved of bone and ivory with a skill and precision unsurpassed until the 1800s.

Remarkable inventions such as these, each a testament to the versatility of the human mind, crowd every page of this fascinating book. As their origins are revealed so too is our unique ability to change and control the world we live in. Many of the origins explored here provoke wonder for their elegant simplicity. Just imagine the delight and surprise of the person who first saw their own image after polishing a piece of the mineral obsidian, and the satisfaction of the folk who found that just hanging meat up to dry in the wind allowed it to be preserved for months. Harder to fathom, however, are revelations such as the fact that by chewing an animal hide Inuit women discovered how to convert it into supple leather.

As well as deliberate design, accident has also played a significant part in the way that everyday things have come into being. According to legend, tea was 'invented' when, more than 4,500 years ago, leaves from a nearby shrub landed in boiling water that the Emperor Chen-nung was about to drink. In a strikingly similar story, silk is believed to have been discovered when, around 2640 BC a silkworm cocoon dropped into some hot water and Xi Lingshi, wife of a Chinese Emperor, saw, as she pulled it out, that it comprised just one single thread. No doubt it was also by accident that the Babylonians discovered that the remnants of a 'barbecue', a mixture of wood ash and animal fat, comprised the essential ingredients of the world's first soap.

As the strands of history from around the world weave through this book clear patterns emerge, making it possible to discern surges of inventiveness such as the Industrial Revolution and the legacies of great explorers that brought us such foods as potatoes and chocolate. The influence of religion is also revealed in the way it has shaped our celebrations – from Christmas, which still retains elements of the pagan midwinter feast of Saturnalia, to Easter, which evolved from the spring festival of the Saxon goddess Eostre.

The Curious History of Everyday Things, as it explores our lives, possessions and customs, is packed with intriguing facts of every kind: that buttons were invented 3,000 years before buttonholes, that the first steam engines were used underground in mines, and that Coca-Cola originally went on sale in 1886 as John Pemberton's 'esteemed Brain Tonic and Intellectual Beverage'. Open this book at any page and you will be sure to find something interesting, informative and surprising.

THE CURIOUS HISTORY
OF EVERYDAY THINGS

THE CURIOUS
HISTORY OF
EVERYDAY THINGS

Contents

CHAPTER 1

HOUSE AND HOME

When the Romans came up with the idea of building multi-storey buildings, divided into individual dwellings, they created the first blocks of flats. Their cities were overcrowded and they were simply responding to a basic need – shelter and security – just as the first humans had when they huddled together in caves or built simple huts. As our early ancestors turned to farming as a way of life they began to live in settled communities and their houses became permanent structures. Since then people have been searching for ways to make a house a home.

Keeping warm has always been a priority. The inventive Romans were the first to develop a central heating system to keep the whole house warm – although modern central heating did not make its appearance until the 19th century. Today's kitchens contain many features that have surprisingly early origins: ovens were being used by the Egyptians around 4,000 years ago and the Chinese had metal cooking pots as early as 1000 BC. Even that apparently most modern convenience, the bathroom, has been around for millennia: in about 2500 BC almost every house at Mohenjo-Daro in the Indus Valley had one. In Britain, however, bathing didn't catch on until late in the 19th century – when Queen Victoria came to the throne in 1837 there wasn't a single bathroom in Buckingham Palace.

From huts to halls

Every room in the house

When the Romans invaded Britain in AD 43, they found that people lived in wattle-and-daub or stone dwellings, with thatch or turf roofs. Inside, families and cattle often shared a single, windowless, all-purpose living room.

LIVING ON ONE LEVEL

City crowding led the Romans to build the first blocks of flats from around AD 64. For similar reasons, flats reappeared in Paris in the 18th century built for middle-class and aristocratic tenants. In London, the Albany building in Piccadilly was converted to apartments in 1803 by the architect Henry Holland. In 19th-century India, however, space was not a problem for the British elite, who transformed traditional Bengali huts – from where the name 'bungalow' is drawn – into much grander single-storey homes.

Roman rulers constructed a series of luxurious country villas and walled cities of sophisticated two-storey brick houses. But after the departure of the Romans in the early 5th century, their materials were pillaged and their consummate building skills were unmatched for hundreds of years.

The emergent Anglo-Saxon civilisation created the first villages and built fine stone churches. Villagers still lived in primitive huts; their lords or thanes in long barn-like halls divided into bays for the family, servants, labourers and livestock.

Multistorey buildings reappeared when Norman barons built imposing stone fortresses at the end of the 11th century. Storage rooms took up the ground floor, with quarters for the garrison and servants on the first. Above them, reflecting the communal lifestyle of early medieval times, was a great hall with steps to private family bedrooms. Rudimentary lavatories were set into the walls of the keep. For most people, indoor toilets would not arrive until after the First World War.

The high cost of transporting stone to the towns meant that timber, and wattle and daub were the main building materials. After a severe fire in London in 1189, Richard I offered incentives to build in stone. A budding merchant class constructed town houses, but the city was still mostly timber-built at the time of the Great Fire of London in 1666.

IN SAFER TIMES

In the countryside, knights lived in fortified manor houses. Early buildings consisted simply of ground-floor storage with a hall or single main room above. By the end of the peaceful 13th century the need for fortifications had faded and the main hall of a manor was often built on the ground floor with adjoining storerooms.

As life became more secure, the lords began to seek more privacy for their families and guests. Ground-floor storerooms were expanded to support an extra room or solar adjoining the main hall. In time, this was known as the parlour, from the French *parler* ('to speak'), or as the drawing room to which people 'withdrew'. Soon this single room was extended into a wing tacked on to the 'upper' end of the hall, behind the head of the table, to create private family chambers.

Just as the hall reached its zenith of splendour in the 15th century, with carved beams and rafters and a large gallery, so its decline began. Its stately height was divided to provide bedchambers or servants' quarters, and less imposing manor houses began to be built. Brick, little used in Britain since Roman times, was reintroduced as a prestige building material during the 16th century.

Although Tudor homes and their rooms bore some resemblance to those of today, they were usually places where people both lived and worked. A town merchant or craftsman would run his business in a shop or workshop on the ground floor below the living quarters; a farming villager might still share his house with his beasts. Home and the workplace gradually began to separate from the 17th century onwards, when rooms such as the parlour and the dining room began to assume their modern functions.

The final knell for the great hall was sounded in around 1651, when Roger Pratt, consulting with his fellow architect Inigo Jones, designed one of the early symmetrical houses, Coleshill in Berkshire, destroyed by fire in 1952. It reflected Italian and French themes with a hall reduced to a central vestibule dividing the building in half, and a double staircase with a flight rising up each side. By the late 1600s, town houses were being built with a hallway rather than a communal hall.

Building blocks
Bricks and stone, mortar and concrete

Around 8000 BC the people of Jericho discovered that wet mud fashioned into blocks would harden in the sun, if it contained sufficient clay. The resulting bricks were their building materials. Some 5,000 years later the Mesopotamians fired bricks in kilns to make them stronger and more water-resistant.

In Britain they were introduced by the Romans: it was common practice to use old Roman bricks – the earliest evidence of new ones dates to the late 1100s. Brick-making machines were developed in Britain from 1825.

Even before brick-making began, stones would have been gathered and piled up to form walls, but hunks of stone were first cut mechanically in ancient Egypt around 2500 BC. In the ancient world clay and bitumen were the most common bonding substances. A mortar containing the mineral gypsum was developed by the Egyptians. At Pompeii, the Romans were building with a mortar of lime, water and sand.

By mixing mortar with pieces of stone and brick, the Romans created a type of concrete. The town walls of Cosa in central Italy, built in 273 BC, are an early example of a Roman concrete construction. From the 1st century AD they made concrete with volcanic sand. In Britain concrete fell out of use after the Romans left until the 18th century.

Windows and doors
Snug and secure

In ancient times the doorway was the only source of daylight in many homes. Some 2,000 years ago, wealthy Romans were the first people to enjoy clear glass windowpanes, produced by casting thin glass blocks and then grinding and polishing them. During the 1st century AD glazed sun porches were a feature of the most luxurious country villas.

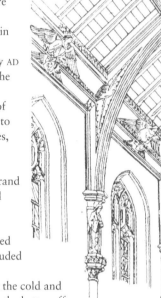

Until the 15th century a variation of the Roman method was the only way to make clear windowpanes, and so they remained expensive and were therefore used only in grand houses and ecclesiastical buildings. For ordinary medieval British homes, waxed parchment or oiled linen stretched over the windows excluded draughts while admitting some light.

The word 'window' derives from the Old Norse *vindauga*, which expressively combines *vindr*, 'wind', and *auga*, 'eye'. For early housebuilders everywhere, climate and vulnerability to attack were the factors critical to window size.

At night, wooden shutters kept out the cold and wind and also gave some security. For the better-off, hinged casements came into use in the 15th century. In these, strips of lead held small panes made of droplets of molten glass poured into moulds.

The art of glass-making was revived in pre-Renaissance Italy via Byzantium. Craftsmen began to spin glass into large discs that were then cut to the required shapes. The first record of this type of glass being used in Britain is for the sash windows installed in the architect Inigo Jones's Banqueting House in London in 1685. In this early type of sash window only the lower frame could be moved. The double hung sash, which opened top and bottom, was in widespread use by the 1750s.

MARVELLOUS MOSAICS

In the 5th century BC artistic Greek builders designed the earliest known decorative mosaic floors, consisting of uncut pebbles pressed into a surface coated with mortar. Colour was first provided by painting the pebbles. The *tessera* technique, whereby stone is cut into tiny geometric shapes that fit in a grid, dates from the late 4th century BC. Glass, first used in about the 3rd century BC, was less suitable than stone for floors.

Throughout Mesopotamia and the ancient world the entrance to a home was closed with hides, cloth or a wickerwork of woven twigs, but doors of this kind were also prevalent. Most Roman homes had wooden doors, while castles and manor houses of Norman Britain were generally sealed with solid oak doors reinforced with metal strips. But until the 16th century most British homes had doors made from soft planks butted together with internal rails and, occasionally, a crosspiece. Stone farmhouses and cottages often had the half-doors now seen mostly in stables. These allowed in daylight while keeping out unwanted animals.

Thatch, tiles and ceilings
Up on the roof

Turfs, moss, and thatch made of straw, leaves, branches or reeds, tied together in bundles, were probably the first roofing materials used in Neolithic homes. In Britain, thatch became favoured everywhere, although it was banned in London in 1212 in favour of stone tiles, as a protection against fire. Tiles of fired clay became economically viable only in the 17th century.

Discoveries in Greece suggest that clay roof tiles date from earlier than 1400 BC. Their use became widespread in Asia and Europe, although wooden tiles were preferred in northern Europe and Russia. The Romans introduced stone tiles, usually made of sandstone and limestone. In Britain high transport costs meant that, until the 1700s, slate was used only for homes near quarries. During the Middle Ages the rafters supporting the roofs of better-quality houses were left exposed; rougher timbers were covered over. In two-storey buildings ceilings were often just the beams and floorboards of the storey above. Decorative plaster ceilings were seen in the 1400s and were common by the 18th century.

Stairways and floors
From top to bottom

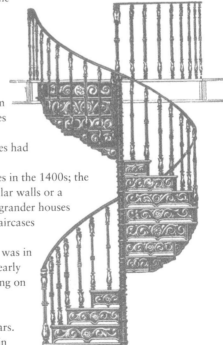

In the first two-storey homes, built in the Middle East from 8000 BC, storeys were linked with simple ladders. Staircases were developed in Egypt in the 2nd millennium BC. Most medieval British homes were single-storey, but tower houses had circular stone staircases built into the walls.

Steep, single-flight staircases appeared in humbler homes in the 1400s; the dog-legged design, with steps built against two perpendicular walls or a chimney stack, followed in the 1500s. By the 1600s some grander houses featured open-well staircases and in the Regency period staircases featuring delicate ironwork were introduced.

Ceramic floor tiles were known in ancient Egypt, but it was in Islamic countries that their use became widespread in the early 1100s. They arrived in Britain via Moorish Spain, appearing on the ground floors of wealthy homes in the 13th century. Stone slabs and floors of beaten earth covered in rushes remained common in ordinary houses for the next 400 years. Wooden flooring for ground-floor rooms was introduced in middle-class homes in the 1700s.

Chairs, stools and sofas
Sitting in style and comfort

From ancient times chairs have been used as symbols of authority, to elevate the mighty above lesser mortals. Some of the first known chairs were created by the Egyptians around 2650 BC for ceremonial occasions: pharaohs and queens were enthroned in ebony armchairs, richly inlaid with gold and precious stones. Medieval lords in England presided over meals from a raised chair at the table while their servants sat in descending order of rank on benches. The word 'chairman', used to denote the leader of an organisation, reflects the traditional link between the chair and status.

The basic form of the common seat has changed little over the centuries. By 1350 BC the Egyptians had made simple square-seated stools, as well as folding wooden stools for domestic and military use. These had X-shaped frames and looked much like those used by campers today.

CLASSIC CHAIRS
The Greeks of around 500 BC were using stylish *klismos* chairs, with simple curved legs balancing concave backs. The Romans knew of them, but developed other types of chairs: one 3rd-century marble relief shows a woman in a basket chair that would not look out of place in a modern bedroom. Likewise, the day bed of the Restoration period and the elegant chaise longue of the 18th century recall the couch beds used by ancient Egyptians, Etruscans and Romans. Even after the fall of the Roman Empire these furniture forms never quite disappeared in the West. They re-emerged strongly when Renaissance craftsmen revived classical models.

Sitting in comfort was a luxury only for the rich who had the time to enjoy it. The settle, a high-backed wooden seat with arms, designed for two or more people, originated in Europe in the 10th century. This developed into the padded settee in the 17th century, when upholstered seating became common. The sofa, larger than the settee and made for reclining, became popular in the 19th century. Its name is derived from the Arabic *suffah*, meaning 'cushion'.

In the early 19th century steel coil springs were fixed within upholstery to provide softer, bouncier seating. The idea of the three-piece suite, comprising two armchairs and a settee, did not catch on until the 1930s when it became stylish, practical furniture for suburban living rooms.

Chests and cupboards
Storage space

For storing bed linen, the Egyptians crafted the earliest know chests from wood or woven reeds in about 3000 BC, while the Minoans of Crete used terracotta for their versions. Wall paintings at Herculaneum, near Naples, created in the 1st century AD, show a fine example of a Roman cupboard, complete with shelves and hinged folding doors.

In Britain the skills for constructing such furniture seem to have been lost after the departure of the Romans, but from the 8th century simple caskets and boxes were made to hold valuables. Later, in the early Middle Ages, larger chests were hollowed from tree trunks and fitted with lids.

By the 12th century, boards were joined to make basic chests for storing valuables or clothes. As the art of carpentry developed, chests were stacked and enclosed in a frame. A 'cupborde with drawing boxes' recorded in a 1596 English inventory was undoubtedly an early chest of drawers.

The word 'cupboard' had first been used in medieval Britain to denote a flat board used for storing cups and plates. Recesses in walls or free-standing pieces of furniture, both with doors, were known as ambries and used for storing food. Medieval Europeans also had armoires, decorated storage units that sometimes incorporated compartments, drawers or, in the case of a Spanish armoire made in 1441, arms on which to hang clothes.

CHIPS OFF THE BLOCK

✤ The art of finishing items of furniture by veneering them with sheets of decorative wood was practised by the ancient Egyptians.

✤ Cofferers, who originally made chests (coffers), and later chairs and desks, were Britain's most important craftsmen in the Middle Ages. When their supremacy was challenged by the import of 'Flaunders chests' from France, the cofferers appealed to Richard III, who imposed a ban on the trade in 1483.

✤ Self-assembly furniture was tried out in the USA during the 1850s but only became successful in the 1960s when Mullard Furniture Industries, or MFI, was founded in Britain.

Board and trestle
The origins of the table

Used, it seems, for bearing sacrificial offerings, the first tables had X-shaped frames and probably looked like oversized stools. The Egyptians built the earliest recorded tables of the ancient world, around 2500 BC, using wood or alabaster. In Mesopotamia tables were made of various metals, while the Greeks used marble.

The Romans produced elaborate small tables, with decoratively inlaid surfaces of cedar, other exotic woods and precious metals, and bronze legs shaped to resemble those of animals. Both Greeks and Romans used their tables for dining and gambling. Large formal tables possibly did not exist in pre-Christian times. Furniture in general was scarce except in the homes of the very rich, and most of it was either built into the basic structure of a house or designed to be easily portable.

The dining table of a medieval lord and his retainers was of trestle construction: oak or elm boards were laid on supports and fixed in place with pegs. After a meal the table could be dismantled and cleared away. The word 'board' meant a table, and this sense is retained in such phrases as 'board and lodging' and 'board of directors'.

Tables with fixed legs joined by stabilising stretchers or tie beams appeared in the 15th century. Over the next 100 years they incorporated modifications such as panels which could be pulled out on runners to extend the table. From the mid 16th century various smaller occasional tables were crafted, including the card table with baize-covered hinged top allowing it to be folded away.

Carpets and rugs
Comfort underfoot

For comfort, furs and animal skins were thrown down on the earth and stone floors of the very first homes. But the oldest complete carpet, found in the Pazyryk valley in central Asia, was made of camel hair and knotted wool. Found in a royal tomb in the Altai Mountains of southern Siberia, it dates from the 5th century BC. However, fragments from Persia (modern Iran) and central Asia suggest that the knotted-pile technique of carpet making was probably developed even earlier by nomadic shepherds, who used their versatile carpets on the ground, as coverings for walls and doorways, and even blankets. Nevertheless, well into the 19th century loose or plaited rushes were still being strewn over the earthen or stone floors of poorer homes, as they had been since antiquity.

The fine handmade linen and wool rugs of Egypt and Syria were brought to Europe in the 12th century as booty from the Crusades. So precious were they that their wealthy owners displayed them on walls or laid them over beds, tables, seating or cupboards.

Royal charters were issued to the carpet weavers of Axminster and Wilton in England in 1701. Aubusson and other smooth-surfaced carpets, which are woven in the same way as tapestries but with a bulkier weft, were the earliest carpets to be machine made, at Kidderminster in 1735.

Decorated walls
Interior style

When, more than 30,000 years ago, the cave painters of the Ardèche in south-east France used combinations of charcoal and earth pigments to make images of big cats, rhinoceroses and bears, they may have been trying to bring good luck in the hunt. But they also became some of the first known interior decorators. By 3000 BC the Egyptians, Sumerians and Assyrians were routinely painting the plaster-covered walls of their homes and palaces using natural pigments from plants and animals.

Magnificent tapestries from Assyria and Babylonia were described by early travellers, but the oldest surviving example is Egyptian. Discovered in the tomb of the pharaoh Thutmose IV, it bears cartouches which contain hieroglyphs describing his life. Hanging painted canvases or other cloths on walls as a cheap alternative to tapestries was also common practice in Egypt at this time, as it became much later in Britain between the 15th and 17th centuries AD.

In the 15th century, wood panelling, originally a Roman technique, was reintroduced into England by Flemish craftsmen. Because it withstood the smoke from fires better than cloth hangings, and also offered some protection against damp weather, it became popular in the great houses of the Elizabethan period. The first wallpapers were decoration for these wood panels: small squares of paper with images printed by wood blocks were stuck onto the panels, then coloured by hand.

Around 1600 shavings of wool (later silk) were blown onto paper covered with adhesive, thus producing flock wallpaper. The technique was already used on cloth. Wallpaper became a less expensive alternative to tapestries and leather wall hangings from the 1680s. Individual sheets were joined together in groups of 12 or more to form a roll, enabling both more complex designs and faster printing. In France a machine for printing wallpaper was invented in 1785. By 1806 rolls of paper were in production in Britain.

HOME EMBELLISHMENTS

✤ Carpet comes from the Latin *carpere*, 'to pluck' or 'to card' (to comb out). It probably derives from the use of unravelled woollen thread for carpet-making.

✤ In 1769 the British designer Edward Beran enclosed wooden slats in a frame to adjust the amount of light let into a room. These became known as venetian blinds from their early use over Italianate windows.

✤ Until they were mass-produced from the late 18th century, ceramic ornaments were confined to the homes of the better-off.

Curtains and shutters
Keeping out draughts

Cloth was such a prized commodity in medieval times that it was seldom used to cover window openings. Curtains were more often used around beds, both to retain warmth and to provide privacy. Plain, heavy cloth, sometimes waxed and enclosed in frames known as fenestrals, might be put in windows to keep out draughts. But it was more common between the 14th and 16th centuries – that is, before windows were glazed – to hang wooden shutters on the inside walls.

The earliest reference to window curtains appears in a 1509 inventory of the nobleman Edmund Dudley's house in Candelwykstrete, London. The inventory mentions 'courteyns of grene … hangyng in the Wynddowys', implying that the curtains were hung from a rod or securely fastened above the window. By the late 16th century the inventories of great houses such as Hardwick Hall in Derbyshire included window curtains suspended from a rod and pulled aside during the day. But curtains were still unusual until the 1700s.

Early window blinds resembled medieval fenestrals, but were made with lighter cotton or silk, usually oiled and painted with decorative patterns. A cloth blind was patented by William Bayley in 1692 and from 1750 springloaded roller blinds, know as 'spring curtains', were available.

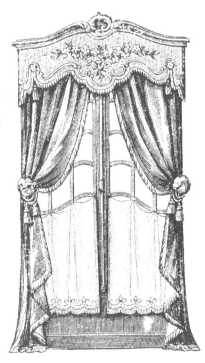

Lamps and candles
Early home lighting

The first artificial light came from open fires, and the flares of burning fat dripping from cooking meat were probably the inspiration for the oil lamp. The earliest known lamps, fashioned by the Palaeolithic cave dwellers of western Europe around 40,000 BC, were hollowed out of stones and contained a wick of moss or other plant material soaked in animal fat. Ancient civilisations developed metal or pottery vessels to burn oil, but the Greeks of the 7th century BC introduced cup-shaped lamps with handles and nozzles or spouts for holding wicks.

Oil lamps changed little through Roman, medieval and Renaissance times. But in 1784 the Swiss chemist Aimé Argand designed a model with a tall glass chimney enclosing a large tubular wick. Draught allowed into the chimney fanned the flame, producing a much brighter light. Lamps burned olive oil, beeswax, fish and whale oil, or tallow, made from the fat of cattle, sheep or horses, until the 1860s when these fuels were superseded by paraffin.

Images of cone-shaped candles were painted on Egyptian tombs in around 3000 BC, and for millennia repeatedly dipping a wick, such as a length of rush or flax fibre, in animal or vegetable fat was the basic method of candlemaking. Beeswax candles, which produced a better quality of light and a sweet, not noxious odour, were too expensive for any but the wealthy.

The invention of moulds by the French nobleman Sieur le Brez around 1600 helped to make candles more plentiful, but poorer families still had to use rushes soaked in tallow. As late as the 18th century, the countryman Gilbert White recorded that 'the careful wife of an industrious Hampshire labourer obtains all her fat for nothing; for she saves the scummings of her bacon pot … A pound of common grease will dip a pound of rushes.'

From gas lamp to light bulb
Illuminating the modern world

The ancient Chinese discovered natural gas while mining for salt, and burned it to create light underground. But the gas lighting that was to take over from the candle and the oil lamp began in France during the 1790s. Gaslights were first installed on London's streets in 1807, and by the 1860s most of Britain's towns and cities were lit by gas.

Early gas lamps burned dimly. Ceiling-mounted lamps with multiple flame fittings were designed to rival candle-burning chandeliers, previously a preserve of the Church and the wealthy, but these 'gasoliers' also produced a tremendous heat.

Competition from the early electric lamps in the 1880s led the Austrian chemist Karl Auer to produce an incandescent gas mantle, made of cotton fabric impregnated with thorium and cerium

AT THE FLICK OF A SWITCH

✻ When electric lights were first installed in hotels and public places in the early 1900s, notices were needed to remind people not to light them with a match.

✻ Electric sockets incorporating on-off switches were introduced by the inventor David Salomons in 1888.

✻ 'If gas lamps are to be superseded, it must be by something that will bear harder knocks than a bulb of glass.' *Boy's Book of Science*, late 1800s.

salts, which gave a brighter, whiter light and used less gas. The cotton burned away leaving behind the metal oxides which were heated to glowing point by the burning gas but did not, themselves, ignite. Light was now so dazzling that it became necessary to shield the eyes from it, leading to the first lampshades.

In 1801 the English chemist Humphry Davy demonstrated that electricity could be used to heat metal strips and make them glow. But it wasn't until 1878 that the English physicist Joseph Swan ran electricity through a carbon filament encased in a glass bulb from which the air had been evacuated, to produce an incandescent light that lasted for several hours. The following year the American inventor Thomas Edison created a light that burned for a couple of days. Honours for the first practical electric light went to Edison.

The origins of the match
The ideal fire-starters

'Rub two pieces of wood against each other to create enough smouldering dust to light dry tinder. Alternatively, strike a flint against steel or other metal such as iron pyrites to create sparks.' So might have read ancient instructions for starting a fire.

In Britain, tinderboxes containing flint, steel and tinder of dry feathers or plant material were essential household items well before Roman times. The flint and steel 'strike-a-light' remained basically unchanged until the 19th century.

Matches of a sort had been known to the Romans; lengths of wood dipped into molten sulphur could be ignited with the heat from smouldering tinder. But even after the Irish scientist Robert Boyle discovered in 1680 that phosphorus and sulphur burst into flame when rubbed together, self-igniting matches still took nearly 150 years to appear. In 1805, French scientist Jean Chancel discovered that wood slivers tipped with potassium chlorate, sugar and gum ignited when dipped into sulphuric acid. But the real precursor of the modern match was the 'friction light' devised by the English pharmacist John Walker in 1827. The head of this match ignited when rubbed against sandpaper.

Ancient utensils and techniques
The prehistoric cook

Cooking probably began hundreds of thousands of years ago, after a piece of raw meat dropped accidentally into a fire. To avoid burnt and shrunken morsels, Neanderthal cooks then took to burying portions of meat in the embers and baking tough vegetable roots on hot stones close to the flames. Carcasses were speared on pointed sticks, suspended between two forked branches and spit-roasted over the fire. Perhaps an early gourmet tried caking meat in mud or wrapping vegetables in leaves to help to retain their succulence.

Nature provided the first cooking pots. Where molluscs or large reptiles such as terrapins existed, their shells made handy heatproof containers. Stomachs and hides of animals were widely used as receptacles and hung over the fire; this method of cooking, used during the 5th century BC by the Scythians, persisted in remote areas of Scotland until the 1700s.

Around 10,000 BC the first clay pots were hand-shaped in Japan; by about 3500 BC the Sumerians of Mesopotamia, who had fashioned cooking pots from stone a few thousand years earlier, were using turntables to throw pots, and utensils such as colanders. Finer sieves for sifting flour in bread-making were made of interwoven reeds, rushes or papyrus.

STRONG AND SMOOTH

The invention of chemical and heat-resistant glass is credited to German Otto Schott in 1884. But it was the American Corning Glass Works that developed the glass cookware Pyrex in 1915. In 1954, Frenchman Marc Grégoire used Teflon, a tough resin discovered 16 years earlier by US company Du Pont, to lubricate fishing tackle. In 1955 Grégoire founded the Tefal company, making nonstick kitchenware.

Metal cooking pots were devised in China around 1000 BC, in the form of the iron wok. Metal pots were known to Britons around the same time. The Romans manufactured saucepans, skillets and frying pans in factories. Most of the cooking vessels now common in kitchens were used in Romano-British homes of the 1st and 2nd centuries AD. One notable exception is the kettle. The medieval 'kittle' was simply a large pan with a lid, used for boiling water or food. Kettles with spouts appeared in the 1690s, but these were luxury items for those who could afford to drink tea. They were made in more affordable materials from the 1720s.

The separate kitchen
A room in which to cook

In hot climates cooking long remained an outdoor activity, but as housing became more sophisticated, food preparation changed its location. The Roman kitchen of the 6th century BC was set in the atrium, the open courtyard at the centre of the house, but by the 1st century AD small rooms containing raised hearth ovens had been tacked onto the back of the house.

In medieval castles and manor houses the kitchen was the centre for all activities requiring heat, including cooking, dyeing and basket weaving. Early kitchens, modelled on monastery halls where mass catering was common, had great open hearth fires and ceiling beams hung with provisions. By the late 1600s, the separate kitchen was an essential part of grander homes.

In the great town houses of the early 1800s, the kitchens were in the basement, run by a battery of staff and rarely visited by the mistress of the house. There might also be a cook's room and a pastry room, situated away from the range. A 'brewhouse' might form part of the scullery, so that the washtubs could be used for making beer.

Revolutions in kitchen design came with advances such as the gas cooker, exhibited at the Great Exhibition of 1851, and the electric cooker, shown in Chicago in 1893. In 1869 Catherine Beecher, sister of the writer Harriet Beecher Stowe, had designed an integrated kitchen including a sink, drainer, worktops and cupboards.

The free-standing cabinet arrived in Britain in the 1920s. One Canadian design consisted of storage space set above a hinged work surface. Seeing the potential, the British manufacturer Len Cooklin established Hygena Cabinets; by the late 1920s the company was producing cabinets equipped with a range of fittings such as a pull-out ironing board.

FOR PREPARING FOOD

✤ The Romans were evidently pastry eaters. They made bronze bun tins and also used pastry crimpers to decorate bread and pies.

✤ Britain's first pastry cutters were created in the 1500s for gingerbread figures. Pastry boards and rolling pins appeared in the 1600s.

✤ The earliest known recipes date from ancient Egypt, written on soft clay tablets that were then baked to preserve them. A papyrus from about 1200 BC lists more than 30 different sorts of bread and cakes.

✤ Before the 18th century, when it came to describe profitable, poor-quality fiction, 'pot-boiler' referred to big smooth stones heated in the fire and thrown into pots to cook food. This ancient method of transferring heat was still used in remote parts of Scotland until the early 1900s.

ALL MOD CONS

Prefabricated standardised kitchen units and worktops were commonplace in American homes by the late 1930s and reached Britain after the Second World War. By 1947 pottery sinks and wooden draining boards were being replaced with sink units pressed from a single sheet of metal and enclosed within a cupboard, that stood directly on the floor. Wipe-clean work surfaces such as the heatproof plastic laminate Formica, developed in the USA in 1913, became standard.

Nostalgia precludes the loss of the kitchen dresser. The original early 17th-century dresser was no more than a board or table where food was prepared or 'dressed' prior to cooking. The single-unit dresser was created by uniting the worktop with the crockery shelves above it, while a cupboard to hold table linen was added underneath. The new dresser became more decorative in function and food preparation was transferred to tables elsewhere.

Fire and fireplace
Taming the flame

Greek mythology recounts that Prometheus stole fire from the gods to give to humankind – and was severely punished for his pains. History suggests, however, that fires were lit in caves occupied by Peking man some 350,000 years ago. Open fires that were contained by stones were the basic pattern for millennia and in the Britain of the 11th century such fires were still occupying the central position of a room.

Only in the 14th century, after chimneys became established, did the fire move permanently to the side of the room. By the 15th century the back wall of the fireplace was protected from the intense heat by a fireback of cast iron. Decorative carved chimney breasts made their debut in grand homes of the 1500s; the mantelpiece, introduced into Britain in Norman times, became fashionable in the late 17th century. Mass-produced cast-iron fireplaces were widely installed during the Victorian building boom.

When coal started to replace wood as the main domestic fuel in the early 18th century, fire grates were designed for burning the new fuel. The new grate incorporated fireback and irons or legs, creating a self-contained unit. From the mid 1700s built-in grates featured hobs, or flat tops on which pots and pans could be heated.

Ovens and stoves
Controlling the heat

More than 25,000 years ago the peoples of the Ukraine circled their hearths with small cylindrical pits, lined them with embers or pebbles hot from the fire, and created the first 'ovens'. Later prehistoric peoples, such as the Neolithic Britons of 5000 BC, coated pits with flat, overlapping stones in order to cook food in water heated with stones from the fire. About 2000 BC the Egyptians devised the first clay ovens.

The Romans fried and grilled food on raised brick hearths. They introduced this technique into Britain, where similar hearths were still in use in the 1830s. The first ranges, produced in the late 18th century, consisted of an open grate flanked by hobs, with an iron oven on one side. Adjusting the ventilation supply to the grate helped to control temperature.

Ranges with insulated enclosed grates were introduced by American-born Benjamin Thompson, better known as Count Rumford. This highly inventive scientist, administrator, philanthropist and sometime spy devised his first closed range for

FIRE TOOLS EVOLVED ALONG WITH THE FIREPLACE. BY THE 15TH CENTURY TONGS TO MOVE LOGS AND A LONG-HANDLED SHOVEL FOR SCOOPING UP ASH WERE COMMONPLACE. THE COAL SCUTTLE AND THE POKER, USED TO BREAK UP LUMPS OF BURNING COAL INTO SMALLER PIECES, WERE INTRODUCED ONLY IN THE 18TH CENTURY AFTER COAL HAD BECOME THE MOST COMMONLY USED FUEL.

the German army in 1798. Its sheet-metal oven was set into brickwork, wells were sunk into the hotplate to hold pots and pans, and beneath each hob was a small fire.

Despite Rumford's advances the enclosed ranges beloved of Victorian Britain were never fuel-efficient or easy to maintain. The survival of the solid-fuel range was ensured by the Swedish engineer and Nobel prize-winner for physics, Gustav Dalen. Blinded in a laboratory accident and confined to his home, in 1924 Dalen created an efficient cooker with an insulated cast-iron firebox, connected to ovens and to hotplates insulated with hinged covers. Produced by the Swedish company Svenska Aktiebolaget Gasaccumulator, from which the name 'Aga' was derived, it arrived in Britain in 1929.

Gas and electric power
New fuels for cooking and heating

Food was probably first cooked with gas at a natural gas spring near Wigan in Lancashire in the mid 17th century. In a letter to the physicist and chemist Robert Boyle in 1687, the Reverend John Clayton describes his experiments with a flame 'so fierce that several strangers have boiled eggs over it' and adds that 30 years earlier it had produced sufficient heat to boil a piece of beef.

This form of energy was destined to be harnessed for the first time in 1802 by the German-born Frederic Albert Winsor (who later brought gas street lighting to London) when he hosted dinner parties cooked by gas at his home in Braunschweig, Germany. Domestic gas fires appeared in the 1850s with the piping of gas into homes. But only when electric-lighting companies made a competitive thrust in the 1880s and 1890s did gas companies begin to improve gas and make it cheaper.

The first patent for an electric heating system was granted in 1887 in the USA to Dr W Leigh Burton. In 1912 Charles Reginald Belling wrapped a filament around a fireproof clay frame to produce a radiant heater similar to those still used today.

In England, St George Lane-Fox designed a means of passing electricity through an insulated wire placed around a cooking vessel. He then patented the first electric 'cooker' device, in 1879. The original electric oven, powered by a waterfall-driven generator, was installed in 1889 in the Hotel Bernina near St Moritz in Switzerland. During the 1890s cooking with electricity began to be promoted but it did not become popular with home cooks until after the late 1930s.

Microwave energy was developed by the British physicists John Randall and Henry Albert Boot for radar installations during the Second World War. It was realised that food could be cooked with it after an exposed bag of maize turned into popcorn. The first microwave oven was patented in October 1945 by the US company Raytheon.

THE FIRST TRUE CENTRAL HEATING SYSTEM WAS THE ROMAN HYPOCAUST. UNDERGROUND FURNACES HEATED AIR, WHICH WAS THEN DIRECTED ALONG TILED FLUES AND DISTRIBUTED UNDER FLOORS AND THROUGH HOLLOW WALLS. MODERN CENTRAL HEATING DATES FROM THE 1800s WHEN MANY LARGER BUILDINGS AND INSTITUTIONS SUCH AS HOSPITALS WERE FITTED WITH CENTRAL BOILERS THAT SENT STEAM THROUGH A NETWORK OF PIPES.

Mrs Lancaster, a writer, eulogised about electricity in 1914: 'It is always willing to do its allotted task and do it perfectly *silently*, swiftly and without mess; never wants a day off, never answers back ... costs nothing when it is not actually doing useful work.'

The origins of the bed

Getting a good night's sleep

Huddled together for protection and warmth, prehistoric hunter-gatherers wrapped themselves in animal skins and slept in shallow pits lined with springy undergrowth. So that they could lie raised above the floor – and any slithering night-life – primitive people planted four forked branches in the ground, built a frame between them and wove thongs together to support a 'mattress' of skins.

From this humble origin the bed became the status symbol of ancient civilisations, and up until the 17th century only the richest people owned one. The ancient Egyptian bed comprised a simple wooden frame and a sleeping surface of interlaced flaxen cords. The Greeks also made their bed frames from wood but wove together leather thongs to create flexible supports.

In medieval England the bed was large and ornate. Reflecting its evolution from the royal tents of military campaigns, it had panelling, curtains and overhead canopies to exclude draughts, drips and vermin. The mighty four-poster developed from this in the 16th century. Elizabeth I used her capacious version as a place from which to conduct state affairs and to meet foreign ambassadors.

The mattress had been developed by comfort-loving Romans. They used a cloth 'envelope' stuffed, according to the owner's economic means, with straw, reeds, wool, feathers or swan's down. In the 6th century BC rose petals were favoured by pleasure-seeking sybarites in

AND SO TO BED

✻ Hammocks, the most portable of beds, were first woven in Central and South America from the bark of the *hamack* tree. Christopher Columbus saw them in Brazil during his voyage to the New World in 1492.

✻ Folding beds, also known as camp beds, were used in French military campaigns of the 15th century. They had elaborate canopies and curtains much like the fixed beds of the day. Napoleon Bonaparte died in a canopied metal camp bed in 1821.

✻ In 1832 the Scottish surgeon Neil Arnott devised water beds as a way of improving patients' comfort.

✻ The poor of the Middle Ages used dried pea pods as mattress fillings, from which the fairy tale 'The Princess and the Pea' may originate.

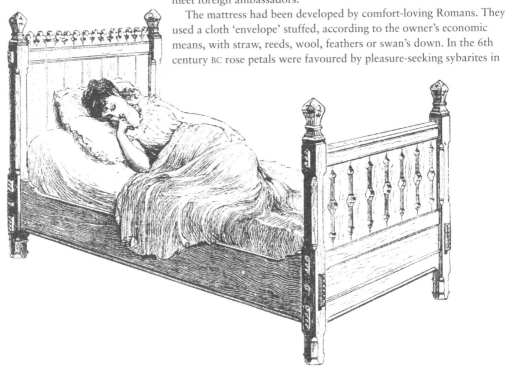

southern Italy, which is why the good things of life may now be described as a 'bed of roses'.

New Yorker James Liddy is reputed to have devised the sprung bed in the mid 1850s when he fell against the seat of a wagon and realised that a network of coiled springs would form a more comfortable bed-base than the latticed ropes used. But in fact, sprung mattresses had appeared in Britain in the 1820s, soon after the spring was first patented.

Comb, brush and mirror
For personal grooming

Some 10,000 years ago Stone Age Scandinavians carved the first combs from bone, while the ancient Egyptians favoured ivory and wood for their double-sided designs. Teeth were cut individually until 1796, when William Bundy, a British textile-machine inventor, developed a device that consisted of several parallel saws. In 1862 the Birmingham chemist Alexander Parkes made the first plastic combs by mixing chloroform and castor oil to make a substance he named Parkesine. This formulation led to the development of celluloid, the material commonly used for combs today.

Compared with combs, brushes for untangling the hair are newcomers to the dressing table, dating back only to the late 18th century. Hogs' bristles from the Russian wild boar were favoured for hairbrushes until supplies dried up after the famine of 1920–21. The backs of the best early brushes were made from ivory, which was imported from Africa or India.

The first mirrors were made around 5000 BC in northern Italy and the Middle East by polishing the lustrous volcanic mineral obsidian. The Venetians of the Middle Ages perfected the process of silvering the back of a sheet of glass, reflecting surfaces were highly polished bronze, tin or silver.

METAL BEDSTEADS WERE MADE IN THE LAST DAYS OF THE ROMAN EMPIRE BUT DID NOT RE-EMERGE UNTIL THE END OF THE 16TH CENTURY IN SICILY. AT A TIME WHEN MANY WOODEN BEDS WERE RIDDEN WITH VERMIN, THEY WERE ADVERTISED IN 18TH-CENTURY FRANCE AS 'BEDBUG-PROOF' ('*NON SUJETS AUX PUNAISES*'). IRON BEDSTEADS BECAME STANDARD IN BRITISH HOSPITALS, PRISONS AND BOARDING SCHOOLS FROM THE 1840s.

Soap and its making
Working up a lather

Ancient Babylonians of about 2800 BC washed their clothes with a form of soap made from wood ash and liquid animal fats. A Mesopotamian recipe from the 3rd millennium BC describes a liquid soap that was made from oil and potash. According to Pliny the Elder, the Phoenicians of about 600 BC cleansed themselves with a soap of goat's tallow and wood ash.

The Romans anointed their skin with oils and hot water, then scraped themselves clean with a strigil. By the early centuries AD, however, they had adopted a Gaulish soap based on the Phoenician recipe. Hard soap was first mentioned at Mount Sapo in Rome, where melted animal fats from sacrificial rites were mixed with wood ashes. Flowing down the mountain they formed a crude soap on the banks of the River Tiber. By the 13th century the Arabs were mixing olive oil with soda ash to make hard soap. Bars were exported from Damascus to Mediterranean cities before soap factories were set up in Castile, Marseille, Genoa and Venice during the 15th century.

A British soap industry emerged in the late 1100s in Bristol, but a tax first levied on it in the 13th century made soap expensive. After the tax was abolished in 1853, the foul-smelling tradition of making soaps at home by boiling up ashes with left-over cooking fat declined rapidly.

FLUSHED WITH PRIDE

In 1449 Thomas Brightfield developed a toilet flushed by water piped from a cistern. But flushing toilets with a plumbed water supply were not common in homes until the late 1800s.

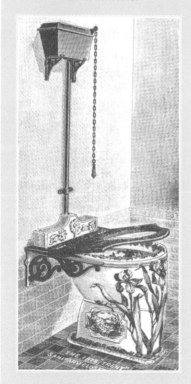

Bathing through the ages
A history of personal hygiene

In the 1930s some flats in Britain were still being built without bathrooms. Yet in the ancient civilised world, cleansing the body with water was considered essential to a person's spiritual and physical well-being.

Around 2500 BC almost every house at Mohenjo-Daro, one of the great cities of the Indus Valley civilisation, had a bathroom. In these little brick structures bathers would douse themselves with water poured from a jug. Waste water was then carried away by chutes or pottery pipes in the floor and into the municipal drains.

Terracotta tubs for bathing were first used in Mesopotamia. Examples of these have been found in the palace at Mari on the River Euphrates in northern Iraq, built about 1700 BC, as well as in the Minoan palace of Knossos in Crete from around the same period. Mohenjo-Daro also had a 'Great Bath' 12m (39ft) long, 7m (23ft) wide and almost 3m (10ft) deep, which appears to have been for communal use. Public bathing in large sunken pools was also favoured by the Egyptians, the Greeks, and especially the Romans, who brought an unrivalled sophistication to bathing and developed efficient plumbing and heating systems throughout their empire.

CLEAN LIVING

From the 11th century monasteries helped to spread a better understanding of the virtues of washing. Monks observed religious rituals involving cleanliness, which were intended to keep them healthier than the general population. Advice came also from the *Salernitan Guide to Health*, produced before 1200 by the renowned medical school at Salerno, in southern Italy. As well as advocating daily washing and bathing every two or three weeks, it introduced the jug and basin, and the wooden bathtub, into the bedrooms of the wealthy. In 1351 Edward III installed the first recorded bathroom in Britain at the Palace of Westminster. It had fire-heated cisterns to provide hot water.

When Queen Victoria came to the throne in 1837 there was not a single bathroom to be found in Buckingham Palace.

As bathing became accessible to the many, poorer people had the luxury of public baths as early as the 12th century. However, protests by the clergy about hygiene and the immorality of mixed bathing brought about their closure. In the 17th century Samuel Pepys's wife was described as having some enthusiasm for public bathhouses, which were a popular meeting place for women. Published in 1640, the *Laws of Gallantry* recommended frequent washing, but despite this Louis XIV of France was said to take only one bath a year.

THE BATHROOM SUITE

In the average home, wooden tubs, which tended to leak and also smelled unpleasant after use, only gave way to lighter metal baths towards the end of the 1700s. Although the Palace of Whitehall contained bathrooms with hot and cold running water in the 1670s, a room used exclusively for washing, complete with its own furnishings, only developed once the technology was available from the 1820s. Victorian catalogues displayed elaborate examples of washbasins, towel rails, showers, toilets, bidets and cabinets with mirrors. Early Victorian showers were hand-pumped, but centuries earlier the ancient Greeks had used rooms with plumbed water that was sprayed on bathers.

Doing the laundry
Keeping clothes clean

From ancient times clothes were washed by being pounded with the hands, or with rocks, in the shallows of a stream or lake. At English well-sides in the Middle Ages, linen was commonly beaten with a paddle-shaped implement called a battledore. In isolated areas of Scotland as late as the 19th century dirty clothes were still soaked in tubs and trampled clean with bare feet.

Commercial laundries were set up by the Romans, but after the collapse of their empire such businesses became a rarity in Britain. Washerwomen and laundresses were commonplace by the 17th century; in the early 1800s larger businesses were laundering for institutions such as workhouses. The first public laundry opened in Manchester in 1842, but such places were a last resort for the respectable. The first US 'Washateria' opened in 1934 and it wasn't until launderettes appeared in Britain in the late 1940s that it become acceptable to wash one's dirty linen in public.

The detergent story
Clean and white

Natron, a natural form of carbonate of soda, was probably the Egyptians' laundry aid. The Romans sprinkled fuller's earth, an absorbent clay, on their woollen garments before beating and washing them clean. This fulling, which also had a thickening effect, was traditionally carried out by men: a Company of Fullers was registered in the City of London in 1376.

The Romans knew that leaving clothes to dry in the sun would lighten them, but they also exploited the bleaching properties of ammonia by soaking garments in a dilute solution of stale urine. In Britain, urine was used as a whitener well into the 19th century, as was dung, which had a similar effect, and lye, an alkali made from wood or plant ashes.

CLEAN BUT DRY

The Mycenaeans of 1600 BC drew out dirt from clothes by sprinkling earth on them then brushing it off. A French book of 1716 mentions the stain-removing properties of turpentine, but dry cleaning was not commercialised until the mid 1800s when the maid of French dyer Jean Baptiste Jolly accidentally spilt turpentine on a piece of cloth.

Until the end of the 18th century, soap was a luxury usually saved for personal hygiene. Clothes might be scrubbed with a harsh soap made from animal fat, but lye was extensively used for boiling laundry. Soda replaced lye from 1791, when sodium sulphate was first produced on an industrial scale from salt by the French chemist Nicolas LeBlanc. The first soap powder, Babbitt's Best Soap, went on sale in New York in 1843. Synthetic soapless detergents, whose molecules attach themselves selectively to dirt, were developed in Germany in the 1880s. Nekal, the first commercial detergent, was marketed there in 1917.

The Romans noticed that adding a faint blue dye to rinsing water masked the yellow tinge of much-laundered cotton and linen and gave the impression of snowy whiteness. The practice was introduced into Britain, along with starch, from Holland in the 1500s, when powdered crystals of cobalt blue or an extract of lapis lazuli were used. It became redundant in the mid 1940s after optical whiteners began to be incorporated into washing powders.

Persil, named from two of its vital ingredients, perborate and silicate, was introduced in Germany in 1907. Persil's secret was self-activation; it 'washed whiter', in the words of their 1930s advertising slogan, by slowly releasing oxygen.

The washing machine
A force for freedom

In the notorious 'Monday wash' of the 19th century, which took at least two days to complete, water was lugged from the well or water-pump and then heated in a giant copper, or boiler, over the fire. Clothes were churned around inside it with a multilegged dolly stick, sometimes known as 'peggy legs', before repeated hand rinsing.

There were many attempts made to ease this arduous job. In 1677 Sir John Hoskins was using a system in which a fine linen bag of his household's laundry was attached to a rotating wheel and immersed in water. William Bailey's hand-cranked device of 1758 reputedly did the work of three washerwomen. In 1858 Hamilton Smith of Pennsylvania patented a machine that consisted of a wooden drum in which a dolly was revolved by hand. Although such machines made washing easier, their tubs still had to be filled and emptied by hand. Real freedom from wash-day drudgery was only possible with the advent of proper plumbing and electricity. The first electric washing machine – the Thor – was introduced in the USA in 1907.

Drying and ironing
To wring and press

From the earliest times, wet clothes were wrung by hand and spread in the sun to dry. Hand-operated mangles and wringers, recorded in Britain from the late 1690s, were standard in middle-class homes of the 19th century. From the 1860s, steam-powered dryers were used in public laundries. The first electric tumble driers were installed in ships of the P&O Line in 1909, but it was the 1930s before domestic electric driers were successful. Until then only well-to-do homes had heated drying rooms and closets – forerunners of the airing cupboard.

Heated irons were used in the Far East from the 8th century, when small pans were filled with hot coals; similar box irons were in use in Europe in the 15th century. By the 1700s the flatiron had become a block of metal, heated on the fire or stove. Gas irons were patented in the 19th century. Although it was developed in France, the earliest patent for an electric iron was taken out by the American Henry Seeley in 1882.

The rise of housework
Keeping a home clean

Epitomised by the clarion call of the Methodist leader John Wesley, 'Cleanliness is, indeed, next to Godliness', a new ethos of purity flowered in the 18th century. Throughout the Industrial Revolution the prosperous middle classes created ever more housework by cluttering their homes with elaborate furnishings and ornaments and by heating their houses with dust-producing coal fires. However, homemade cleaning materials, such as bone ash and charcoal polish, dating from about the 17th century, remained in popular use.

Until the early 1900s the cleaning ritual began at dawn with raking out the kitchen range, then scrubbing it clean with a blacking solution and polishing it. Tea leaves or sand, scattered over stone and tile floors, helped to dampen down dust before sweeping, washing and scrubbing. Tea leaves or freshly pulled grass were also scattered on carpets before they were brushed clean; occasionally they were also taken outside to be beaten free of dust.

A FEATHERED FRIEND

The feather duster, that icon of housework, may have an even more ancient origin than the besom, for brushes made of feathers were almost certainly used in around 25,000 BC in the execution of cave paintings in northern Spain. In Britain, moss was another natural material employed for dusting, certainly until the 18th century.

Vacuum cleaners and carpet sweepers
Gobbling up the grime

In 1901 the London engineer Hubert Cecil Booth placed a handkerchief on an upholstered seat and proceeded to suck air through it. By thus depositing grime on his handkerchief, he proved that suction could dislodge and trap dirt. Inspired by an unimpressive demonstration of a dust-blowing machine for cleaning railway carriages, Booth went on to build the first machine to combine a power-driven suction pump with dust-collecting. This cumbersome horse-drawn device, from which a long hose would be run into a building, was given the nickname of 'Puffing Billy'.

Smaller, bellows-operated vacuum cleaners evolved for domestic use, but were eclipsed by the 'electric suction sweeper' patented in the USA by James

Murray Spangler, an asthmatic school caretaker, in 1907. Unable to capitalise on his invention, he sold it to a leather-goods maker, William Henry Hoover, whose original upright, bag-on-stick model of 1926 was marketed with the slogan: 'It beats as it sweeps as it cleans.'

Brooms and brushes
A clean sweep

When people first kept house, 10,000 and more years ago, dirt was swept up with branches plucked from shrubs. So the brush was 'invented', and then improved by binding flexible sticks and twigs together around a wooden handle. In Europe, favourite plants for this purpose were broom, from which the brush took its name, heather, birch and the tufts of maize plants. The first American brush factory was founded in New York in 1859.

Brushes of bristles knotted together in bunches were made by the Romans, but not until the 15th century were hogs' bristles glued with pitch into blocks of wood and used for scrubbing and for brushing clothes, as well as for sweeping. In the USA the shape of the broom was altered, and its efficiency tremendously increased, when in the late 18th century the members of the Shaker religious community, with elegant simplicity, converted its circular head into a flat wedge.

Washing-up and dishwashers
Tackling the dirty dishes

During the Middle Ages, when chunks of stale bread were used as 'plates', washing-up was non-existent. Even after wooden platters came into use they were generally simply wiped clean with bread or straw. It was the widespread introduction of china, following the expansion of the Staffordshire potteries at the end of the 18th century, that heralded the chore of washing-up.

The dishwasher was invented in 1885 by Eugène Daquin, a Frenchman. His machine used a revolving mechanism equipped with artificial hands that grabbed dishes and dunked them first in soapy water, then in a clear cold bath; rotating brushes scrubbed the dishes clean. A report on the machine in *Scientific American* assured readers that the dishwasher constituted 'no danger whatsoever to man or dish'.

In 1886 American Josephine Cochrane patented a small domestic dishwasher. Inspired by the number of plates her servants broke, she designed a machine in which a hand crank sprayed soapy water over the dirty dishes. The first dishwasher to be electrically powered appeared in 1922.

THE HYGIENIC HOME

❉ The Romans used weasels, not cats, to rid their homes of rodents. By the Middle Ages various kinds of mousetrap existed, including a metal contraption with a large spring and serrated jaws, illustrated in a German book of the 13th century.

❉ A paste of sand, grit and ashes was the usual 17th-century scourer. Victorian housewives used washing soda (sodium carbonate), but this was eclipsed by Brillo steel-wool pads, first sold in the USA in 1913 with an accompanying bar of soap.

❉ Until the 1920s most people made their own cleaning products. From Roman times, furniture polish was made from beeswax mixed with juniper or cedar oil.

❉ Rubber gloves first protected hands from hot water in the early 1950s.

Safes and money boxes
A place to hide valuables

In ancient times people concealed their valuables in caves or up trees, or simply buried them, but in Europe wooden chests were used to provide security for jewels and other precious objects from the 8th century AD. Strongboxes of wrought iron, strengthened with interlaced iron bands, were being made in large numbers in southern Germany by the late 16th century, in sizes ranging from a few centimetres to about 1.5m (5ft) long. Used throughout Europe, they became known in Britain in the mid 19th century as armada chests, in the mistaken belief that such chests had held the bullion that financed the Spanish Armada in 1588, or that they came from wrecked Spanish ships.

THE KEY TO HAPPINESS

In ancient Rome brides were ritually presented with the keys to their new households, and adulthood is often still marked with receiving 'the key of the door'. The symbolic power of the key can be traced to Greek mythology, in which it unlocked the door to Heaven and Hell. Its association with good luck, and the superstition that a key should be kept in the keyhole to prevent evil spirits from entering the home, originate from ancient Mediterranean culture.

SAFE AND SOUND

In the coffers used by the wealthy of the 17th and 18th centuries to secure their riches, an elaborately engraved lock usually occupied the whole interior of the lid. The keyhole on the front of these early safes was usually a decoy, the real one being concealed beneath a panel on the lid.

In the early 19th century various metal safes were patented, including several by Jeremiah Chubb. However, most of them were intended to be fireproof rather than burglarproof. The first patent for a fireproof safe was granted in 1834 to the Englishman William Marr. Safes continued to be simply variations on the theme of the iron chest until the much publicised Cornhill robbery of 1865, after which they were more finely engineered.

The idea for a time lock, which opens only at a predetermined hour and runs by clockwork or electricity, was conceived in 1831 by the Scottish bank agent William Rutherford. These came into their own in the USA in the 1870s when burglars began torturing bankers to force them to hand over the strongroom keys or reveal the combination.

The penny-wise have long found ways to hide their savings. Hollowed flints filled with Roman coins from the 1st century AD have been unearthed in southern England. The first containers made specifically for keeping coins were used in ancient Egypt and Rome to store offerings to household gods. Onion-shaped pottery money boxes with a slit for inserting coins were used by apprentices in medieval Britain to collect Christmas tips from their masters' customers. On Boxing Day the pots were broken to release their bounty. Animal figures date from the 17th century. The pig, which gave us the 'piggy bank', was one of the most popular figures, probably because of the importance of pigs in rural economies.

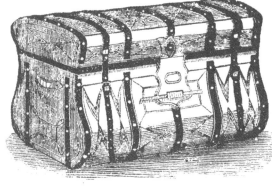

Protecting property
Under lock and key

The need to keep thieves at bay is as old as the notion of property. Simple locks in which a wooden bolt was attached to a piece of rope, pulled into place through a hole in the door and secured with a complex knot, were known to the Chinese and the Egyptians as early as 2000 BC. The first tumbler lock, named after the falling pins that secured the bolt, was in use in Iraq during the 8th century BC. Its subsequent popularity in Egypt, where it was used for securing mummy cases, led to it being named the Egyptian lock, and it spread through Greece to the rest of Europe.

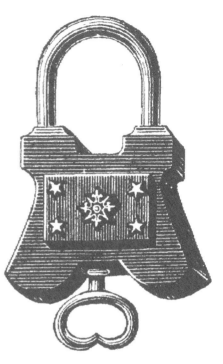

By the Middle Ages elaborate locks were made throughout Europe, but they were easily picked. The first major design improvements came from Britain. In 1778, locksmith Robert Barron patented the double-action tumbler lock, based on the principle by which mortise locks operate. The tumblers were released by a first turn of the key, and the lock opened by a further turn. Some 40 years later, Jeremiah Chubb added a detector tumbler to jam the lock if anyone attempted to pick it. To prove the principle, in 1818 a Chubb lock was given to an imprisoned locksmith-turned-burglar with a promise of £100 and a pardon if he could pick it. After trying unsuccessfully for over two months, the frustrated convict gave up.

Joseph Bramah was a cabinet-maker who became one of the most inventive engineers of his age. In 1784 he devised a lock operated by a complex cylindrical key containing slits and notches that lowered sprung segments in the lock. He was so sure of his lock that in the early 1800s he offered a reward to whoever could pick it. The challenge was finally met in 1851 by the American Alfred Charles Hobbs, who took 51 hours, spread over 16 days, to perform the feat.

The American locksmith Linus Yale made a lock based on the Egyptian principle of pin tumblers. Today's Yale lock was developed from this by Linus Yale Junior in 1856 and has proved a lasting success.

BOLT FROM THE BLUE

Both the ancient Egyptians and the Minoans of Crete erected what seem to have been lightning rods on temples, although their exact function is unproven. Lightning remained a puzzle until 1752, when the American statesman and inventor Benjamin Franklin proved that it was a form of electricity by flying a kite into a cloud during a storm and showing that a key attached to the kite had become electrified. His subsequent invention of the lightning conductor earned him the venom of the clergy, who accused him of interfering with the wrath of God.

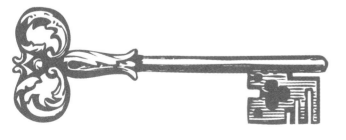

The pedigree of dogs
Man's best friend

From cosseted Pekingese to doleful bloodhounds, all our canine pets are descended from wolves. The ancestor of the wolf itself was a carnivorous tree-climbing mammal *Miacis*, which lived around 40 million years ago. Distinct canines first appeared 35 million years ago in North America.

Research by biologists in the United States suggests that dogs have been domesticated for some 135,000 years – a process that started before modern humans began to leave Africa 100,000 years ago. But archaeological evidence of dogs goes back only to 12,000 BC, the date of a Palestinian grave. Buried dog remains have also been found in a Stone Age settlement in Yorkshire.

Domesticated dogs were trained to chase, bring down and retrieve prey, and to guard camps and other settlements. A greyhound-type dog was depicted chasing gazelle on a piece of Mesopotamian pottery, 8,000 years old. The Romans bred dogs for specific tasks, importing different types from all over the empire, and credited them with watchfulness, courage and loyalty. They were also the first to issue the warning 'Beware of the dog' (*cave canem*), which can be seen beneath a fierce-looking canine on a Pompeiian mosaic dating from the 1st century AD.

PRIZED PETS

✣ In ancient Egypt, 'Mau' was an affectionate name given to cats. Dead cats were lovingly embalmed before being given ceremonial burials. In one grave, at a site of a temple to the cat-goddess Bastet, 300,000 mummified cats were discovered.

✣ The first widely reported dog show was held in 1859 at Newcastle upon Tyne. The world-renowned Cruft's Show, named after its founder, Charles Cruft, began in 1891.

✣ The earliest book to set guidelines for judging dogs was published in 1867 and listed 35 British breeds.

✣ Isaac Newton, the English physicist, is said to have invented the cat flap in the late 17th century.

ROYAL FAVOURITES

The Egyptians kept the first lap dogs, which they depicted sitting under the chairs of their owners. Such dogs were popular in China, where crossbreeding and strict diets to impair growth produced small dogs with bent legs and snub noses. Pekingese have existed for 2,000 years: originally they were the exclusive property of the Chinese imperial court.

Spaniels, whose name probably comes from the Old French *espaignol* (Spanish dog), are first described in a French book on hunting from the 14th century. They became fashionable some 200 years later when Henry III of France kept miniature spaniels. The King Charles spaniel is thought to have originated in either China or Japan, but earned its name in 17th-century England from Charles II's passion for his pet spaniels.

The first poodles were large gun dogs used by hunters to catch waterfowl – hence their German name, *Pudelhund*, meaning 'splashing dog'. French or German in origin, they were already known in the 13th century. Traditionally, the poodles' thick, curly coats were clipped to make them resemble lions. This also allowed them to swim more easily, while the bands of hair that were left around their legs and front quarters protected their joints and vital organs from the cold.

Feline pets
The domesticated cat

Compared with their canine rivals, cats were late converts to domesticity. The earliest known remains of a tamed species, with teeth significantly different from those of Mediterranean wildcats, were uncovered in Khirokitia in Cyprus and date from around 6000 BC.

Cats were first depicted – hunting wildfowl, killing mice and playing under chairs – in Egyptian art around 1500 BC. They were domesticated when it was noticed that they protected granaries from rats. Domestic Egyptian cats may have reached Europe on board ships plying the trade routes to Phoenicia.

The association of cats and devil worship took hold in the 12th century. In the 13th century the Church launched a crusade against two French heretical sects who were said to have worshipped the devil in the form of a black cat, and the religious persecution of cats began. Europe's cat population plummeted. When plague struck in the 14th century there were not enough cats to kill the rats that were the main carriers of the disease-spreading fleas: a quarter of Europe's human population perished.

Cats regained their domestic niche in the 17th century, when they were regularly portrayed in Dutch paintings. A century later they were beloved pets in many British households. Imported breeds were introduced from the 1880s, including the first Siamese cats ever to be exhibited, given to Owen Gould, a British consul general in Bangkok, by the king of Siam (now Thailand). Like Persians, Siamese cats are thought to be descended from an ancient species of Asian wildcat.

EGYPTIAN CAT OWNERS WOULD SHAVE THEIR EYEBROWS AS A SIGN OF MOURNING FOR THEIR PET CAT. THEY TOOK THE BODY TO BUBASTIS, CITY OF THE ELEGANT CAT-HEADED GODDESS BASTET, ALSO KNOWN AS BAST OR PASHT, WHO WAS SAID TO POSSESS NINE LIVES. SHE MAY ALSO BE THE DERIVATION OF THE AFFECTIONATE TERM 'PUSS'.

Caged birds
Popular winged pets

The ancient Egyptians delighted in the exotic birds that surrounded them and depicted at least 70 different species in their art. The lapwing, the hoopoe, the golden oriole and the crane were among those birds that were kept in captivity.

The canary, named after its native Canary Islands, was domesticated in the 1500s. The budgerigar was brought to Britain from Australia in 1840 by the naturalist John Gould. Its name comes from the Aboriginal *budgeri*, 'good', and *gar*, 'cockatoo'. The breeding and showing of budgerigars began in earnest in the 1920s. The affectionate term 'budgie' dates from the 1930s.

Gadgets for the home

'Every new invention that is practical and economical…will be brought to [your] notice…The time spent on housework can be enormously reduced in every home, without any loss to its comfort, and often with a great increase to its well-being…'

'THE REASON FOR GOOD HOUSEKEEPING', *GOOD HOUSEKEEPING* MAGAZINE, MARCH 1922

For nearly every domestic chore there is, and many other tasks besides, a gadget has been invented somewhere. The process started early: as far back as the 1st century AD the Greek engineer Hero of Alexandria designed a tiny steam turbine for opening doors. In the 9th century the three Banu Musa brothers of Baghdad wrote *The Book of Ingenious Devices*, which included an oil lamp with an automatically rising wick. Roasting spits operated by the heat rising from a fire instead of by human spit-turners were known in the Middle Ages.

As the machine age took hold in the 19th century, so labour-saving machines multiplied. Among the more practical was the American Universal boot cleaner of 1882. Its hand-turned wheel had a row of bristles for brushing off dirt and another one for polishing.

Kitchen aids

Mechanical whisks, food mixers and mincers appeared in the 1850s. By the 1880s the classic mincer design had emerged from the London firm of Spong and Co. It had a hand-turned screw to drive the food through fixed blades, as well as various attachments for slicing vegetables and meat. Smaller versions were sold for use at the dining table by those who had lost their teeth.

An earlier British invention was the coffee grinder, which came into use in the late 1600s. Until then spices and coffee had been ground using pestles and mortars. Small grinders for home use were first made in 1815 by the iron founder Archibald Kendrick. His much copied

coffee mill consisted of a bowl at the top to hold the beans, a handle to turn the blades and a drawer into which the ground beans fell.

Canned food, which was being produced from the early 1800s onwards, demanded a means of getting at its contents. A can of veal made of tin plate and taken on an Arctic expedition in 1824 bore the instructions: 'Cut round on the top with a chisel and hammer.' In the USA, pierce-and-prise can openers were introduced around 1858. Simple cutting wheels were used from 1870; only in 1931 did a design with pivoted handles and a turning key appear.

A pierce-and-prise can opener was one of the features of the pocket knife designed by Karl Elsener for the Swiss Army in 1891, in which a spring mechanism allowed utensils to be folded next to the main blade. The Swiss Army knife, initially issued only to recruits, was so popular that by 1900 it became available to civilians.

Power tools

In the late 19th century inventors rushed to make use of electricity. The first electric toaster, the Eclipse, was sold by the Crompton company, of Chelmsford in Essex, in 1893. Its bare wires, glowing red hot as the current passed through them, toasted bread one side at a time. The pop-up toaster, which used a clockwork timer to turn off the electric current and release a spring under the toast, was devised by Charles Strite, a mechanic from Stillwater, Minnesota, in 1927. The Proctor toaster of 1930 improved on this by using a thermostat to 'read' the surface temperature of the bread, a principle that is still employed in today's models.

The electric food mixer, introduced in the USA in 1910, was transformed by the British engineer Ken Wood, whose first Kenwood Chef was marketed in 1950. Cooks could now mix, shred, liquidise, mince, and even open cans, simply by fixing on one of a variety of tools.

Early electrical gadgets were not confined to kitchen use. In the late 1880s the French hairdresser Alexandre Godefroy produced a gas hair dryer: this alarming device took the form of a bonnet fitted with a flexible tube that was connected to a gas stove. By 1899 an electric hand-held dryer was available in Germany. The electric toothbrush, first patented in the USA in 1885, did not achieve commercial success until a practical design was made by the Squibb company of New York in 1961. Even the job of getting a perfect crease in a pair trousers was mechanised: in the 1950s, John Corby of Windsor added a pressing device to his popular valet stand and so created that favourite of hotel rooms – the Corby trouser press.

Scissors and shears
Cutting things up

Among the handiest of all household objects, scissors first appeared in the Bronze Age. However, their origins go back more than a million years, when clubs made of antlers and hammers of stone were first employed as domestic tools, and when sharpened stones were mankind's only cutting implements.

Bronze, first made in Mesopotamia about 3500 BC, provided the material for better blades. The first scissors, in widespread use in Europe and Asia by 1000 BC, consisted of two blades connected at the handle end by a C-shaped spring. This design, which was used to cut everything from hair to animal hides, remained basically unchanged for millennia, and was being used for sheepshearing until recent times.

Scissors inspire many superstitions, one of the most common being that accepting a pair as a present from a friend risks severing the friendship. Traditionally a small coin is given in exchange for the gift.

In the Middle Ages the nobility of Europe had made elaborately decorated gold or silver spring scissors inlaid with enamel, pearl and diamonds. Well into the 17th century humbler folk, and craftsmen such as tailors and seamstresses, used simpler versions made of either iron or brass.

Pivoted scissors, in which a screw or rivet axis connects two blades with handles looped at their ends to hold the fingers, were being made of bronze and iron 2,000 years ago in Roman Europe and the Far East. More expensive to make than spring scissors, they did not become common for domestic use until the 16th century. Large-scale production began in 1761, when the Sheffield metalworker Robert Hinchcliffe began to make them from cast steel, which produced stronger blades.

Nails, screws and bolts
Pinning things down

The hammer, originally just a roughly shaped stone, has changed little since the Egyptians produced the first cast copper hammerhead around 3500 BC. Claw hammers, with a V-shaped opening opposite the head for extracting nails, appeared in Roman times.

Nails are equally ancient. The oldest known were found on a statue of a bull, consisting of copper sheets nailed to a wooden frame, made in Mesopotamia in about 3500 BC. But in the Middle Ages nails were still being made by craftsmen who hammered rods through a succession of holes of decreasing size, and flattened the end to make a head. Consequently they were expensive and used only to fix metal to wood. Most joiners and builders fastened timbers together with cheap wood joints and pegs. The first nail-making machines were developed by Ezekiel Reed in the USA in 1786 and by Thomas Clifford in England in 1790.

Screw-type devices such as the water screw for raising water were known to the Egyptians. But the idea was applied to carpentry only in the 16th century, when it was realised that a nail with a twist in its stem stayed more securely in place. A slot was cut into the head to make it easier to fit and remove, and a 'turnscrew' designed. The first machine for cutting screws was patented by Job and William Wyatt in England in 1760.

Nuts and bolts appeared in about 1550, together with the spanners needed to tighten and undo the bolts. Like early screws, they were hand-made and of slightly different sizes. A screw-cutting lathe developed by Henry Maudslay in 1797 standardised production, making pieces interchangeable. Around 1700 the adjustable spanner appeared in France.

INNOVATIONS ROUND-UP

✻ Rubber bands were first made by Perry and Co of London in 1845.

✻ Charles Gould invented the stapler in 1868 for use in bookbinding.

✻ A vacuum flask was demonstrated in 1892 by the Scottish physicist James Dewar. His mother-in-law remained sceptical of a flask that Dewar made for his son in 1902 and knitted a woollen cosy for it to make sure that its contents stayed warm.

✻ Vacuum flasks were first marketed in 1902 by the German Reinhold Burger, who held a competition to name the invention. The winning entry was Thermos, the Greek word for 'hot'.

✻ In 1900 the Norwegian Johann Vaaler patented the simple paperclip.

✻ The aerosol can was invented in 1926 by another Norwegian, Erik Rotheim, but the idea was not developed commercially until 1941, when Lyle Goodhue and William Sullivan took out a patent in the USA, and used the spray can for insecticides.

Adhesives and glues
Sticking things together

Excavations at Umm el Tlel in Syria have revealed that some 40,000 years ago, early humans attached stone tools to wooden hafts with bitumen, a naturally occurring dark, sticky substance now used for surfacing roads and waterproofing roofs. For the ancient Sumerians of around 5000 BC, bitumen, oozing its way through cracks in the earth, was associated with the underworld and evil spirits, a belief that may be the origin of the Biblical Hell's 'lake of pitch'. The practical waterproofing qualities of bitumen were first exploited in about 2400 BC, when shipwrights in Babylon and Mesopotamia used it to caulk their ships.

The Egyptians manufactured glues by boiling animal skin, bone and sinew, and similar glues are still used by traditional carpenters. Other natural adhesives used since earliest times include beeswax, egg white, gum, resin and starch pastes. Synthetic glues, developed in the 20th century, are stronger and more versatile.

'Superglues', discovered accidentally by scientists of the American company Eastman Kodak in the 1950s and first sold in Britain in the mid 1970s, provide the tightest grip of all. Investigating the light-bending qualities of the substance ethyl cyanoacrylate, the scientists were surprised when the glass prisms of the refractometer stuck firmly together and could not be prised apart.

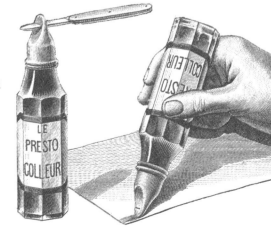

Sacred blooms
The power of flowers to inspire

Throughout history, flowers have been offered as gifts to the gods or endowed with divine significance. The fabled lily of Mictlan, the lily of the valley, is an Aztec legend. Wanting a flower for their heavenly gardens, the gods sent a bat to steal a tear from Xochiquetzel, the goddess of trees and flowers. The tear was then transformed into a lily that had the gift of eternal bloom, but it was both colourless and without fragrance.

So the gods sent the bat with the lily to Mictlan, the underworld, to be washed in the River of Death. There the lily received the gift of beauty but lost that of eternal bloom. Rejected by the gods since its beauty was doomed to fade, the lily finally took root on Earth, where for a brief season it has silver blooms and a wonderful fragrance. The lily is the property of Mictlan and its beauty is only on loan to the Earth.

In ancient Egypt, the water lily, also known as the sacred lotus, was used in arrangements to decorate holy places. Vases dating from 2500 BC, found in the tombs of the necropolis Beni Hasan, beside the River Nile, were put in place to hold the scented blue Nile lotus. The lotus also appears in early Indian Buddhist and Chinese Taoist scriptures and art.

Two thousand years ago, the fragrant sweet violet, which grew in the Mediterranean region, was linked to Aphrodite, the Greek goddess of love. It became one of Europe's favourite flowers, brought into homes for its scent and simple beauty.

Floral decoration
The art of arranging flowers

FROM THE 1950s CENTRAL HEATING AND DOUBLE GLAZING MADE EUROPEAN HOMES SUITABLE HABITATS FOR A WIDE RANGE OF EXOTIC PLANTS, INCLUDING THE AMARYLLIS, NATIVE TO SOUTH AMERICA, AZALEAS FROM ASIA AND THE AFRICAN VIOLET. THESE ALL ADDED TROPICAL SPLENDOUR TO LIVING ROOMS AND CONSERVATORIES.

Displaying flowers in the home is as important now as it was in the 2nd century AD when cut flowers artfully arranged in a container were depicted in a mosaic for the emperor Hadrian's villa in Rome. Ikebana, the Japanese art of formal flower arrangement, began in the early 7th century and is based on Buddhist philosophy.

Flower arranging was not regarded as an art form in Europe until the 17th century, when exotic new flowers were imported from discovered lands. The lavish bouquets of baroque paintings gave way in the 18th century to a neo-Classical revival of Graeco-Roman floral tributes. In British homes, 'bough-pots' were used to arrange branches and flowers in the hearth in the summer months. The mass-production of goods made possible by the Industrial Revolution brought cheaper ceramic and glass containers, making flower arranging more popular.

ALOE

Potted plants

Growing exotics indoors

Gone to Pot

❊ Succulents such as the American aloe or agave were introduced to British gardeners from the Americas in the early 18th century. Cacti were rarely grown outside botanical gardens until the 19th century.

❊ Plants were grown in terracotta pots and window boxes in the crowded urban communities of ancient Rome. Large, decorative plant pots, known as jardinières from the French *jardin*, 'garden', became popular in England in the 1800s.

❊ The popularity of cut flowers was revived in the late 1800s as electricity replaced gas lamps.

The practice of house-plant cultivation began in earnest in Europe only after the late 1600s. In 1653 the agricultural authority Sir Hugh Plat published *The Garden of Eden*, a reissue of a 1608 title, *Floraes Paradise*. It was the first English guide to growing plants indoors. Soon after, the wealthy began to add greenhouses and conservatories to their homes. Modest householders also looked to exotic plants to brighten the gloomy parlours of 18th and 19th-century homes. Plants that could withstand the fumes from gas lighting became popular, despite their lack of flowers. One example was the aspidistra, dubbed the cast-iron plant because of its durability and resistance to the smoke and dust of open fires. Its successful transition from Chinese forest undergrowth to Victorian parlour was possible because both its fleshy roots and its leathery leaves were naturally adapted to intermittent rainfall as well as to deep shade. The Victorian fashion for palms was greatly encouraged by the opening of the Palm House in Kew Gardens near London in 1848. The exotic Canary Island date palm and fan palms became favourites and were considered sophisticated additions to any conservatory.

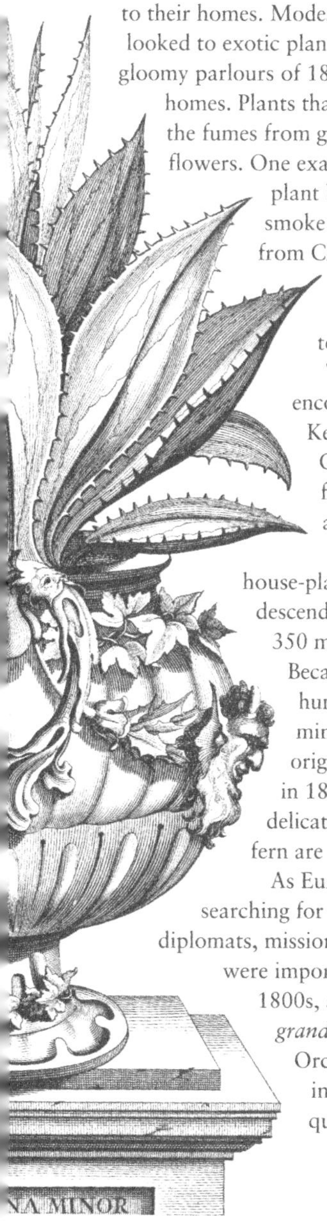

One of the most passionate of the Victorian house-plant crazes centred on ferns, the modern descendants of giant swamp forest plants that thrived 350 million years ago in the Carboniferous Age. Because many ferns demand high levels of humidity, they were grown in Wardian cases, miniature table-top greenhouses named after their originator Nathaniel Ward, who invented the case in 1829. Of the ferns that survived the craze, the delicate ribbon brake and the tongue-like bird's-nest fern are still popular.

As European empires expanded, plant hunters searching for plant seeds followed in the footsteps of diplomats, missionaries and travellers. North and South America were important areas of botanical exploration in the 1800s, and many species, including azaleas, *Begonia grandis* and lilies, were also brought from China. Orchids from tropical forests arrived in the West in Victorian times and their cultivation indoors quickly developed into 'orchidomania'.

Earthly paradise

'And the Lord God planted a garden eastward, in Eden…
And out of the ground made the Lord God to grow every
tree that is pleasant to the sight, and good for food…'

GENESIS, CH. 2, V. 8-9

The perfect garden, an enclosed space with water, trees and flowers, is an ideal that has travelled from the origins of Western culture into towns, cities and suburbs around the world. Irrigating and protecting crops and other useful plants from the elements and from potential marauders, were the first motivations for gardening. But even plants grown for the table can be objects of delight, and the pleasures of the garden were quickly appreciated.

Garden glories

No records exist of gardens before the 2nd millennium BC, but Egyptian tomb paintings of about 1475 BC depict temple gardens, and the pharaohs brought plants from Africa, Europe and Asia for their gardens by the Nile. In about 1200 BC Rameses III made 'great vineyards; walks shaded by all kinds of sweet fruit trees, a sacred way splendid with flowers from all countries, with lotus and papyrus as numerous as the sand'.

Conquests and commerce carried the cultural development of the Egyptian garden to Persia, where emperors built intimate pleasure gardens full of shade and water, large enclosed game reserves and terraced parks planted with trees and shrubs. *Pairi-daeza*, the ancient Persian name for such gardens or parks, is the root of the word 'paradise'. The elaborate Hanging Gardens of Babylon, built in arid lands now in central Iraq by Nebuchadnezzar II during the 6th century BC, were one of the Seven Wonders of the Ancient World. The ancient Greek writer Diodorus recorded that 'the several parts of the structure rose from one another tier upon tier'.

Courtyard gardens of urban Greece were lined with colonnades, while at public sports grounds philosophers would stroll through groves of cypress and olive, fir and poplar. As the empire expanded, ideas from the East were employed to create luxurious orchards and vegetable gardens – and the 'smiling patch of never failing green' encountered by Odysseus on a visit to Alcinous, king of the Phaeacians.

The Romans adopted Hellenistic traditions but further adapted the garden paradise to the empire's diverse climatic conditions. In Italy's towns and cities villa courtyards were outdoor 'rooms', often with landscapes painted on the walls to create an illusion of space. Fruit trees, flowers and vegetables flourished in this sheltered environment. In the country the great villa gardens were parks planted with orchards and vineyards, and with decorative trees and shrubs. When the Romans reached Britain they brought with them lilies, as well as figs, mulberries and peaches.

A thing of beauty

Gardening in the Middle Ages became less flamboyant, but medieval monastic and castle gardens grew plants for their medicinal, cosmetic and culinary uses. The illuminated manuscripts and tapestries of that time also show plants clearly appreciated for their beauty. By the 16th century flowers had become cherished for their appearance and scent, and many of today's suburban garden features began to emerge.

Ball games of all kinds were loved by the Tudor nobility, who needed flat areas of short grass for play, and the lawn evolved from the sports alleys and greens of 16th-century England. Many early lawns, made from meadow turf, were floral masterpieces of violets, daisies and speedwell. By the 1650s the bowling green had become distinctive to English cultivation, with meadow flowers outlawed as weeds. Garden neatness was seen as a virtue from the 18th century, when the lawn became a neutral green foil for flowering plants in borders. A passion developed for lawns manicured to appear, in the words of the 19th-century novelist Anthony Trollope, 'as much like velvet as grass has ever been made to look'.

The water of life

In arid climates water, the essential element that made gardens possible, became central to their design. In the garden oases of ancient Egypt, Persia, the Mediterranean and northern India, irrigation technology progressed to produce fountains and waterfalls for geometric water gardens. During the 1st century BC the Roman architect Vitruvius described the Mediterranean water-garden tradition, which later provided the inspiration for Italian Renaissance architects. Their feats of engineering produced the renowned Italianate water gardens.

A new style emerged in Britain in the 1700s, led by the landscape designer Lancelot 'Capability' Brown, who got his name from his habit of describing gardens as having 'capabilities of improvement'. As a result, formal water gardens were replaced with naturalistic lakes and pools.

A world of flowers

Hunting out new species

In the Middle Ages, gardens featured little but roses, lilies, pinks, cowslips, marigolds and violets, but by the end of the 16th century more than 200 kinds of plant were grown in England, including tulips, anemones, hyacinths and crocuses. In the following century lupins, phlox, goldenrod, Michaelmas daisies and Virginia creeper were all brought from North America. By the late 1830s, there were 18,000 species of garden plants grown in British gardens, with more emerging yearly from both the Orient and the Americas.

The spread of printing and new written works from the late 15th century led to greater access to information about plants. Between the 16th and 19th centuries, the increasing availability of new species from around the world, the rise of nurseries and the publication of gardening books introduced flower cultivation to a wider public.

EXOTIC HARVEST

From Mexico came the dahlia, introduced into Europe in the 1790s. In the early 1800s the Empress Josephine, wife of Napoleon I, developed dahlia hybrids and varieties. These later became part of her fantastic arrangements of exotics in the gardens of Malmaison, outside Paris.

Many of the flowers introduced into Europe were brought back by British 'plant explorers'. In the early 1600s, on expeditions financed by his employers, the Cecils, earls of Salisbury, John Tradescant Senior collected plants from across Europe and the Mediterranean region, including the gladiolus, rock roses and the lilac. John Tradescant Junior also brought back many new perennials and shrubs from his own journeys to North America.

BOTANICAL BOUNTY

Under the patronage of George III, Sir Joseph Banks, who had documented the flora of Australia on his travels with Captain James Cook, was made the first 'director' of Kew Gardens in 1771. Following this, and the foundation in 1804 of the future Royal Horticultural Society, many expeditions were initiated. The explorations of the Scottish botanist David Douglas in California in the mid 1820s led to the introduction of, among other species, the flowering currant, the California poppy and the monkey flower.

Rhododendrons and azaleas were the bounty of 19th-century botanical expeditions such as Sir Joseph Dalton Hooker's journeys to the Himalayas and Ernest Wilson's adventures in China. Wilson collected plants for the Veitch family, who ran one of the most influential nurseries of the 19th century and who made possible the introduction of countless plants. More than 400 of them were illustrated in William Curtis's *Botanical Magazine*. Founded in 1787, it was the first gardening magazine and is still published under the title *Curtis's Botanical Magazine*.

The romance of the rose
The story of a symbolic flower

The rose is red, it is said, because it blushed with pleasure when it was kissed by Eve in the Garden of Eden. In ancient Egyptian and Greek cultures the rose was linked to female deities: in Egypt to Isis, in Greece to Aphrodite. The Greeks also used rose petals to cure anyone bitten by a mad dog.

In Britain only a small selection of roses was grown before the reign of Elizabeth I. Nevertheless, one of the earliest uses of the rose as an English symbol occurred during the Wars of the Roses in the 15th century. The white rose of York was probably *Rosa alba*, a native of northern Europe that had been introduced by the Romans. The red rose of Lancaster was *Rosa gallica* var. *officinalis*, the old rose of Provins in central France. It was reputedly brought to Europe from the Middle East by Thibault IV, ruler of Navarre, during the Crusades. Later it became known as the apothecary's rose because of its common use in ointments and potions.

From the 1500s rose breeding gained in popularity. The moss rose, *Rosa centifolia* 'Muscosa', a variant of the cabbage rose, appeared in a catalogue of 1727. Chinese roses were being grown in Kew Gardens by 1769. The first of the tea roses, *Rosa odorata*, also from China, was introduced by Sir Abraham Hume in 1810, but it was the plant collector John Parks who began to market them in the 1820s. Their scent was said to resemble that of fresh-packed tea.

Tulips and other bulbs
Fortune in a flower

Along with other flowering plants, bulbs, corms and tubers made their first appearance on Earth about 35 million years ago. Most are native to the Middle East and Mediterranean, and to places with similar climates, such as South Africa. Irises were grown in classical times, and lilies, irises and daffodils all adorned the Islamic gardens of the 7th century. Tulips, anemones and crown imperials were also nurtured in the gardens of the sultans.

In the 1550s Ghiselin de Busbecq, the Holy Roman Empire's ambassador to the sultan in Constantinople, sent the bulbs he had seen growing in Turkish gardens back to Vienna. Here, they were received by the botanist Carolus Clusius, who in 1594 took his tulip collection to the Low Countries.

Tulips had first reached Holland during the 1570s. By the beginning of the 17th century 'tulipomania' had taken hold, and Dutch merchants were gambling large sums of money on single bulbs in the hopes of breeding flowers with streaked petals or 'breaks'. In the early 1620s one bulb of the Semper Augustus tulip was sold for 20,000 stuivers – this at a time when the annual wage of a skilled carpenter was just 8,000 stuivers. Before the crash in 1637, another example of the same tulip was said to have sold for 260,000 stuivers.

In the 1920s, science revealed that the variegated tulip petals, which commanded such huge sums from 17th-century Dutch merchants, did not result from the skill of plant breeders but were caused by a virus transmitted by aphids.

Garden tools
Home horticulture

Digging, weeding, watering, pruning and other gardening tasks have changed little through the ages. The plethora of hoes, spades, shovels, rakes, trowels, forks and shears listed in an inventory at Abingdon Abbey, Oxfordshire, in the 14th century still comprise the basic set of gardening tools.

Flint hoes with wooden handles were made in Mesopotamia around 4000 BC and both the ancient Egyptians and Romans used iron hoes. These were pulled towards the body – the technique still favoured by British gardeners today. Until the 19th century, short-handled hoes, which were used in a kneeling position, remained more common than long-handled designs.

Load-carrying shovels were made in the Stone Age from the shoulder blades of large food animals such as deer, and from these evolved the spade for digging. Early spades, probably devised by the Romans, were scoops fashioned from wood with an iron rim attached. Alternatively, a wooden handle was inserted into an iron cutting blade. The Romans also used the *trulla*, or trowel.

Shears based on the same principle as Roman spring scissors were first used for sheepshearing, but shears for hedge trimming, lawn edging and pruning evolved from the 16th century with the introduction of pivoted scissors. Secateurs, with curved blades, had been developed in France by 1818 and proved useful for the new rose hybrids. These, unlike their predecessors, required careful pruning to ensure blooming in subsequent years. Named from the French for 'cutters', secateurs quickly became popular in Britain. In 1830 steel shears made by the Sheffield company Steers and Wilkinson were praised for their strength. 'By using both hands,' reported one tester, 'the most delicate person may cut through a branch of an inch in diameter.'

MOWING MACHINES

Lawns became the focus of suburban gardening in the early 19th century and in 1830 Edwin Budding found a way to make light of lawn care. Budding, a Gloucestershire engineer, saw a machine used in the textile industry for shearing the nap from woollen cloth and realised that it could be applied to grass. Previously grass had been mown by hand with scythes or cutters pulled by donkeys. Ransomes, an agricultural machinery company, was manufacturing Budding's lawn mower by 1832. In 1902 Ransomes produced the first petrol-driven motor mower to be commercially successful.

Before specialised equipment was developed, gardeners used a wide range of devices for carrying water to their plants, including ladles and gourds. Early watering pots of the 1470s were made of clay. They had no spout but many small holes in the body that released fine streams of water. Spouted pots were first depicted in the 1600s, with the term 'watering can' appearing in 1692. Copper watering cans with handles replaced clay pots from about 1700.

Garden hoses made from tubes of fabric originated in the 1400s, but the lawn sprinkler was made practicable only through the invention of vulcanised rubber hosing in the 1860s. In 1871 the American J Lessler of Buffalo, New York, patented the garden sprinkler.

CHAPTER 2

MANNERS AND CUSTOMS

Children dressing up as ghosts and ghouls on Halloween may seem like a relatively recent custom but this celebration of the supernatural has ancient origins. October 31st was once Samhain Eve, the night before the beginning of the Celtic new year when the souls of the dead were believed to roam the Earth once more and the festival of Samhain the following day was celebrated with purifying bonfires.

Many other traditional celebrations are also adaptations of ancient rituals. Valentine's Day has its roots in the bawdy celebrations of the ancient Romans at the festival of Lupercalia on 15th February, while dancing round the maypole recalls a pagan fertility ritual. Equally, many superstitions grew out of early man's attempts to control or appease the frightening power of Nature and are often linked to religious beliefs. Crossing fingers for good luck resembles the sign of the cross made by early Christians and the number three was considered lucky by the Anglo Saxons who used it to break spells.

However, the trappings of some customs are more modern. White weddings, Christmas trees and Mother's Day cards are largely Victorian inventions although the traditions they commemorate have been observed for centuries.

Codes of politeness
Common courtesies

With arms outstretched, displaying hands empty of weapons, early humans approached their neighbours to offer or ask for help. In these or similar gestures of peace and friendship the notion of etiquette was born. As societies developed, so too did codes of conduct. Position in society played a significant role in matters of etiquette, a word derived from the French for 'label'. Social conventions often emanated from the royal courts, with monarchs devising courtesies applied selectively according to rank.

The earliest reference in English literature to a strictly stratified society regulated by a code of conduct that marked the distinctions between classes and ranks appears in the 10th-century Anglo-Saxon epic poem *Beowulf*. It describes how Queen Wealtheow offers to drink a toast to the king and then to his courtiers in a clearly defined order of precedence.

BAD BEHAVIOUR

The V-sign, with the palm facing inwards, may have its origins in war. It is said that on the eve of the Battle of Agincourt in 1415 the French threatened to chop off the 'bow fingers' (first and second fingers) of every English archer. At dawn next day the victorious English extended the same two fingers in a mocking gesture. During the Second World War a V-sign, with the palm facing out, was used by Winston Churchill, the British Prime Minister, as a victory symbol.

EARLY ETIQUETTE

In the West, the tradition of a code of conduct is intimately linked with the Church and early medieval chivalry. The *Pax Dei* (*Peace of God*) was a set of principles adopted during the 10th century for Christian warriors. It was instrumental in the evolution of the ideal of chivalry, which flourished between the 11th and the 13th centuries during the Crusades.

Throughout the Middle Ages rules that related to morals and manners were often communicated through popular songs and rhyming verses. Some examples appear in the English 15th-century *The Babees' Book*, for pages and maids-in-waiting. In one piece, entitled 'Urbanitatis' ('Of Politeness'), *The Babees' Book* instructs them:

'When you come before a lord
In hall, in bower, or at board,
You must doff or cap or hood …'.

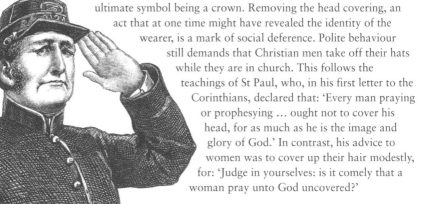

Historically, the covering of a person's head symbolises authority – the ultimate symbol being a crown. Removing the head covering, an act that at one time might have revealed the identity of the wearer, is a mark of social deference. Polite behaviour still demands that Christian men take off their hats while they are in church. This follows the teachings of St Paul, who, in his first letter to the Corinthians, declared that: 'Every man praying or prophesying … ought not to cover his head, for as much as he is the image and glory of God.' In contrast, his advice to women was to cover up their hair modestly, for: 'Judge in yourselves: is it comely that a woman pray unto God uncovered?'

The handshake and the kiss
Bonding behaviour

A hand signal can be seen before a face comes into focus – so we greet with a wave, then smile to demonstrate our welcome. To show that they came in unarmed friendship, the Romans greeted each other with outstretched arm and open palm. In the 20th century this salute took on a more sinister meaning when adopted by Nazis under the dictator Adolf Hitler.

In ancient civilisations, such as the Kingdom of Israel and the Roman Empire, the handshake signified the binding of a contract. In medieval times it marked a pledge of honour or allegiance; a subordinate would kneel while his superior initiated the handclasp. Shaking hands as a polite form of greeting across social classes, although considered too egalitarian to be acceptable in earlier ages, became acceptable in early 19th-century Europe.

If, as anthropologists argue, the loving kiss is related to sucking in babyhood, it is as old as the human race. The less intimate kiss on the cheek was used as a form of greeting or salutation among the ancient Greeks and Romans. The Hebrew Old Testament also records that 'Esau ran to meet him [his brother, Jacob] and fell on his neck, and kissed him'. Subsequently, kissing became a symbol of Christian brotherhood: 'Salute one another with a holy kiss,' St Paul urges in his Epistle to the Romans. Medieval lords would ceremonially kiss newly invested knights on the cheek. The knights, however, showed respectful submission or allegiance to their lord by kissing his hand.

Establishing decorum
Guiding manners

The Dutch scholar Desiderius Erasmus believed that most people not only aspired to gentility but also wanted to know the rules of behaviour. His short treatise *De civilitate morum puerilium* (*On the Civility of the Behaviour of Boys*) of 1530, covered manners at school, in church, at mealtimes and even in bed. It also included discreet guidelines on how to 'name some functions of the body that our sensibility no longer permits us to discuss in public'.

In France from the mid 17th to the early 18th century Louis XIV insisted that courtiers observe an elaborate timetable of civilities. Dandies such as Beau Nash and Beau Brummell set the rules of British etiquette during the 18th and early 19th centuries. Their whims swiftly became the fashion. The Prince Regent (later George IV) is said to have wept when his friend Brummell disapproved of the way his coat fitted.

In the United States, George Washington advised on good behaviour in his pamphlet *Rules of Conduct*. In the 19th century, 'How to Choose and Win Friends' was covered by 'Miss Manners', the journalist Judith Martin. America's most celebrated expert on etiquette was Emily Post, whose 1922 book *Etiquette: the Blue Book of Social Usage* became an instant best seller.

THE TERMS U (UPPER-CLASS) AND NON-U (NOT UPPER-CLASS), COINED IN 1954 BY THE BRITISH LINGUIST ALAN ROSS, BECAME WIDELY KNOWN IN 1956 THROUGH *NOBLESSE OBLIGE*, A COLLECTION OF ESSAYS EDITED BY THE SOCIAL COMMENTATOR NANCY MITFORD. WORDS SUCH AS 'PUDDING' WERE DEEMED TO BE 'U' WHILE 'SWEET' WAS 'NON-U'.

Feasting together
Sharing meals

When food was scarce, our early ancestors, the hunter-gatherers, had no regular breaks for eating. Mealtimes began to be adopted when Neolithic people started to control their food supply through agriculture and animal domestication, and settled down. The taking of meals reflects the rituals, status, religion and etiquette of specific cultures. Sharing food marked covenants, or promises, in both the Old and New Testaments. Today birth, marriage and death, as well as religious celebrations such as Passover and Christmas, are celebrated in meals.

In Western societies breakfast, lunch and dinner form the standard daily pattern of meal-taking, although local variations abound. The timing of meals and their content are dictated by working hours and tradition, as well as by the availability of foodstuffs. In 13th-century Europe, for instance, the largely agricultural population ate two main meals each day. The first was dinner at about 9 or 10am, and then later there was supper at 5pm. Meals would consist typically of grain-based broth or cereal porridge. This was accompanied by onions and green vegetables such as cabbage, or bread. Except for the nobility and wealthy farmers, meat was rare. Only in the early 20th century in North America did 'dinner' become exclusive to the evening.

EATING ETIQUETTE

Table manners are markers of society and culture. To eat with a knife and fork, first established as polite in the 17th century, is typically Western, while for many Muslims the convention is to use the right hand. Pronouncements on etiquette have featured in almost every age. *The Babees' Book* of 15th-century manners for young people advised:
'When you are set before the meat,
Fair and honestly it eat.
First, look ye that your hands be clean ... '.

BREAKING THE FAST

In the Middle Ages religious orders adhered to a daily two-meal regime, while the working classes took short breaks for food during the day, the first of which would 'break the fast' of the previous night. By the 17th century, however, breakfast had become a substantial meal, consisting of cold meats, fish, cheese, bread, butter and wine or ale for the rich, and cereal pottages or porridge for the poor. The traditional British breakfast dishes – milky oat porridge or eggs – have rural roots, since both milk and eggs could be supplied fresh every morning after the farmer had attended to his cattle and poultry. In the 18th century, when an increase in urbanisation meant that many working men ate chops and steaks in the newly established chophouses, rather than coming home from their jobs in the middle of the day, dinner was customarily at 5pm. The resultant gap between breakfast and dinner gradually led to the institution of a midday snack, which by the 19th century had stretched to become an extensive meal called luncheon, eventually shortened, both in duration and verbally, to simply 'lunch'.

As dinner became a later meal, eaten by the upper classes at around 7.30 to 9pm, 'afternoon tea' evolved from a simple hot drink to a meal in its own right. At first, nursery teas were for children, who were sent to bed before dinner. In the mid 19th century taking tea became fashionable among middle-class and aristocratic women – who were possibly following the example of Anna, Duchess of Bedford – in an effort to legitimise the secret habit of snacking on cakes in their boudoirs.

The roots of friendship lie in the sharing of food, and the word 'companion' stems directly from the Latin *com panis*, 'with bread'.

Some families continue to dine early, and even today 'high tea' can be a substitute for, rather than an addition to, dinner. However, in some households, 'dinner' can also mean the midday meal.

Dishes, plates and cutlery
Different kinds of tableware

Prehistoric diners ate off large leaves and flat pieces of wood, while for the Romans, pastry cases were served as both dish and food, rather like pizzas today. In medieval Europe, people ate off 'trenchers' – slices of stale bread which soaked up the juices and might later be fed to the dogs or given to the poor. A fresco dating from 1525 in the town of Mantua, Italy, depicts some of the earliest recorded flat metal plates, and by the end of the 17th century flat ceramic plates were common in France. In Britain a 'service' of matching plates, bowls and serving dishes became affordable only after Josiah Wedgwood developed mass-produced earthenware in the 18th century.

The first cutlery consisted of simple 'spoons' and scoops made of shells, hollowed-out pieces of wood and socket joint bones. When people first learnt to bake, clay spoons were among the earliest implements to be made. Egyptian spoons were bronze, and often had a pointed handle to extract the meat from snail shells. Wealthy Greek and Roman families used silver spoons. However, it was not until Georgian times that the knife, fork and spoon appeared together as eating implements on British tables.

ANCIENT LUXURIES
Most cutlery now in everyday use was once a luxury available only to the wealthy. Even in Europe's aristocratic circles, diners shared drinking cups until the 16th century, and guests also brought their own knives to the table.

In Britain the same applied to forks until the late 17th century. They were probably introduced in the early 17th century by Thomas Coryat, who described forks in his travel book *Coryat's Crudities*. Before that, diners speared meat with a knife-point or used their fingers. Early table forks had two prongs; in the 1800s three and four-pronged table forks were used.

The knife has a far longer history, founded on cutting materials of prehistoric times such as flint, obsidian, bone and shell. The art of slotting flints into wood and serrating the edges was refined by the Egyptians in the 4th millennium BC. By 1500 BC a variety of bronze cutting tools was being honed from Britain to China. Rich Egyptian and Roman families owned ornamented table knives, and the Romans also devised knives that had steel blades.

Taverns, inns and restaurants
Eating in public

In the taverns of ancient Rome, slaves and other lowly members of society ate hunks of cheese washed down with cheap Cretan wine. Here people exchanged the gossip of the day. The taverns of early medieval Europe, which had evolved from the Roman *tabernae*, were also places frequented by the poor and were devoted mainly to the consumption of alcohol and simple foods. Such was their unsavoury reputation in Britain, however, that in the 8th century the Archbishop of York decreed that 'No priest [should] go to eat or drink in a tavern'. Inns, where lodgings were provided for the increasing numbers of travelling merchants, were also popular in the Middle Ages. From the 16th century a set menu for dinner at a fixed price came to be called an 'ordinary' or *table d'hôte*, meaning 'host's fare'. The inns where it was available soon became known as 'ordinaries' too.

Samuel Johnson, the 18th-century writer, lauded the tavern: ' ... there is nothing which has yet been contrived by man, by which so much happiness is produced as by a good tavern or inn.'

FINE DINING

As food quality improved, these ancestors of today's restaurant became convivial meeting places for businessmen. Until the 18th century in Britain, dining out had been predominantly for the poor, with the wealthy preferring to entertain in style in their own homes. The first restaurant to be described as such was opened in Paris in 1765 by a Monsieur Boulanger. Until then, only traders known as *traiteurs* were allowed to sell cooked meats, which people were not permitted to consume in public.

During the 19th century grand hotels, many of them incorporating excellent dining establishments and employing highly gifted chefs, grew to cater for the expanding middle classes. The renowned French chef Georges-Auguste Escoffier worked with the Swiss hotelier César Ritz at the Savoy in London and at the Hotel Ritz in Paris in the late 1800s. Escoffier did much to establish modern dining principles, shortening menus and speeding up service. He also courted his wealthy clientele, for instance by naming new dishes in their honour. Pêche Melba, a dessert of poached peach, ice cream and raspberry purée, was created for Dame Nellie Melba, the famous Australian soprano.

Food phrases and conventions
Language and traditions

Like the restaurant itself, much of the modern language of dining comes from France, including the words '*chef*', '*maître d'hôtel*', a head steward or head waiter, and '*entrée*' – which in French means an 'entry', or starter, but which in the 1900s came to mean a main course in English. *Hors-d'oeuvres* were dishes placed 'outside the work': originally these appetisers were grouped around, or outside, the culinary centrepieces laid out for a banquet.

The French word '*menu*', from the Latin for 'small', acquired its sense of 'detailed' and then 'list' during the 19th century, appearing as the *menu de repas*, the 'meal list'. To 'wait', which came to be used for serving food at the table, emerged in the 1500s from the Germanic word *wahtan*, meaning 'to watch'. And *à la carte* dining – the 'card' being the menu – was introduced as an alternative to fixed *table d'hôte* meals after 1765, when Boulanger's restaurant was opened.

Since many of the staff in early restaurants had previously worked in the households of the nobility, their servants' outfits were smartly adapted to become the familiar black-and-white uniform of the waiter. The tradition of giving a gratuity to a waiter originated in 16th-century Europe from the popular practice of making small payments to boys for running errands. During the late 19th century European workers took the custom with them when they emigrated to the United States.

> THE TERM 'TIP' PROBABLY CAME FROM COFFEE-HOUSES IN THE CITY OF LONDON, WHERE MERCHANTS GATHERED TO CONDUCT BUSINESS IN THE 17TH AND 18TH CENTURIES. MESSAGES FOR SWIFT DELIVERY WERE PLACED IN A BOX WITH A SMALL COIN AS PAYMENT, THE LID OF THE BOX BEARING THE LETTERS T.I.P. – TO INSURE PROMPTITUDE.

Fast food
Snacks and speedy meals

Nearly 4,000 years ago, street vendors in Ur in Mesopotamia sold fried fish, roast meat and onions. More recently, Vienna sausages and frankfurters were first sold as popular fast food in the USA from the early 19th century. The 'hot dog' gained its name in 1906 when the American cartoonist Tad Dorgan drew a dachshund (the 'sausage dog') in an elongated bun.

The fast-food boom that followed the Second World War was epitomised by the McDonald's chain. Richard and Maurice, the McDonald brothers, established a small drive-in restaurant in Pasadena, California, in 1937 selling hot dogs. By the 1940s hamburgers had begun to replace hot dogs, but not until 1948, at their San Bernardino outlet, did the McDonalds standardise low-cost food served at speed in clean surroundings.

But it was Ray Kroc, a salesman for electric milk-shake mixers, who turned McDonald's into an international brand. In 1954 Kroc delivered an order to the San Bernardino branch. He was so impressed that he bought the rights to franchise the McDonald concept. In 1955 the first of the new restaurants opened, in Des Plaines, Illinois; Britain's first dates from 1974, in London.

Valentine's day
Anniversary of love

Spring has long been considered the season of love. In ancient Rome revellers at the festival of Lupercalia on February 15 celebrated fertility with bawdy bands of near-naked youths running around the streets. Around the 4th century AD the Church initiated the festival of St Valentine's Day on February 14 to replace such pagan excesses. It appears that neither of two possible Valentines, both 3rd-century Christian martyrs, had any connection with romantic love. The hint of impending springtime, when animals usually mate, simply served to confirm the natural link with Lupercalia.

From the 13th century a set of customs developed around the festival. Women who were not partnered could be won in the Valentine lottery, a Lupercalian throwback which survived until the 19th century. Luck, accident and anonymity characterised the spirit of the festival: it was a tradition that the first person seen on February 14 would be one's love for the rest of the year and Valentine cards remain unsigned today.

LOVE TOKENS

Red roses were first given as a St Valentine's Day gift in 18th-century France, in imitation of Louis XVI's love tokens to his queen, Marie Antoinette, although these flowers had symbolised heavenly perfection and earthly passion since ancient times. Credit for sending the first valentine often goes to Charles, duc d'Orléans, who conveyed to his wife 'poetical or amorous addresses', which he called *valentiens*, while he was imprisoned in the Tower of London after the Battle of Agincourt in 1415. However, the modern valentine, the earliest form of greetings card, evolved during the 16th century. Early valentines were handmade and decorated with knots, hearts, doves and cupids. In composing messages for their cards, lovers could turn for inspiration to *Valentine Writers*, a handbook of standard rhymes and greetings referred to in 1723. The first printed valentine dates from the mid 18th century.

Subsequently, the practice of sending valentines grew more popular, and by the 19th century postmen were claiming a special meal allowance for the extra load of mail. Around this time a variety of decorative styles emerged, including mechanical valentines with figures that rolled their eyes and put out their tongues, and cards adorned with cupids that hovered or rowed boats when the cards were opened.

STRAIGHT FROM THE HEART

The death of John Balliol of Barnard Castle, the wealthy 13th-century landowner and founder of Balliol College, Oxford, possibly gave rise to the use of the word 'sweetheart' as a romantic term of endearment. Throughout the remainder of her life, Balliol's widow, Devorgilla, devotedly carried a silver and ivory casket containing her beloved husband's embalmed heart. When she died in 1290 the casket was buried beside her by the high altar of the Cistercian abbey in Dumfries she had founded just 17 years earlier. Henceforth it came to be known as Sweetheart Abbey.

Becoming engaged
Pledges of betrothal

Betrothal was once much more than a verbal promise. In Roman times bridegrooms of the upper classes had to provide security for the completion of the bargain, while the early Church required a contract of betrothal to be sworn in front of witnesses. If either party died during the engagement, the survivor inherited a share of the other's estate. Among poorer people the rules were more relaxed and trial marriages were common. The Roman law of *usus*, or 'prescription', specified that if a woman lived with a man for a year without being absent for three nights they were considered married.

Asking for a woman's hand on bended knee is one of the lasting legacies of chivalry and dates back to the Crusades when European knights were influenced by Arab culture. With her acceptance came the giving of a ring. Only much later, in Elizabethan times, did the wedding ring begin to assume a similar importance. By the 19th century the Church had changed betrothals into mere announcements of a wedding, and engagement rings became less important than the ring required for the matrimonial ceremony.

In Greece and Rome it was believed that a vein ran from the third finger of the left hand directly to the heart. This may well be why an engagement ring was traditionally placed on this finger.

The dating game
Making a match

The world's first official marriage bureau was established in London in 1650. In its prospectus, founder Henry Robinson confidently announced: 'Such as desire to dispose of themselves or friends in marriage, may here likewise be informed what encounters there are to be had.'

On July 19, 1693, the earliest known British matrimonial advertisement appeared in the publication entitled *Collection for the Improvement of Husbandry and Trade*, announcing: 'A Gentleman about 30 Years of Age that says he has a Very Good Estate, would willingly Match Himself to some young Gentlewoman that has a fortune of £3,000 or thereabouts. And he will make settlement to Content.'

In 1958 two Americans, Shirley Sanders and Robert Kardell, became the first couple to be matched by computer, and were introduced to one another on a television show, *People are Funny*. They were married that October in a Hollywood church and given a Honolulu honeymoon by the show's sponsors.

Tying the knot

'We will have rings, and things, and fine array; And, kiss me, Kate, we will be married o'Sunday.'

THE TAMING OF THE SHREW, ACT II, SC. I, WILLIAM SHAKESPEARE, 1593–94

The institution of marriage evolved in the first human societies to protect lines of descent and ensure that children were safely brought up. Well before 2000 BC the Sumerians had established laws concerning marriage, although free choice of spouse was not a feature of matrimony until modern times. Roman law was the first to introduce the requirement of consent to a union by both the bride and the groom.

Wife capture, practised among the early tribal Britons, proved unpopular with the Anglo-Saxon settlers of the 5th century, since women were a valuable family asset, spinning flax and wool for cloth – hence the word 'spinster'. Consequently, in the 6th century marriage by purchase became law. Family elders fixed the bride price and the term 'wed' arose, from the Old English for a pledge.

At first Anglo-Saxons only visited the early Christian priests to have their weddings blessed, but, by the time of the Norman Conquest, what had once been purely a business alliance had become a religious rite. In the 16th century the Church decreed that every impending marriage should be announced by the public reading of banns, and that the wedding should be celebrated in church. This revived an early Christian custom: the 2nd-century theologian Tertullian noted that in Carthaginian society all marriages were considered clandestine and illicit unless announced in church.

Marriage vows

In medieval Britain a costly alternative to public banns was available in the form of a special licence permitting a couple to wed privately and without having to announce the event in advance. The public calling of banns became a legal requirement with the English Marriage Act of 1754. Civil law retained the option of special licences, and these were considered more genteel.

The ring has proved an abiding symbol of marital fidelity, although other tokens, such as a halved coin, were common in

Britain until the 17th century. In Sumeria around 10,000 years ago rings probably evolved from the shackles placed on brides to subdue them. The Greeks were among the first to inscribe rings with sentiments, such as 'mayest thou live'. And although the plain band of gold or silver has proved the most enduring design, there have been many alternatives, such as two hearts joined by a key, popular in Roman times.

Placing the ring on the third finger of the left hand originated with the ancient Greeks and Romans. From the 11th century in church ceremonies a priest placed the ring over the tip of the index finger 'in the name of the Father', then over the second 'in the name of the Son', before lodging it on the third finger 'in the name of the Holy Ghost'. But wedding rings were often worn on other fingers, and on the thumb, even in the late 1700s.

Anne of Brittany married the French king Louis XII in 1499 wearing the first pure white wedding dress on record. By the end of the 16th century wearing white symbolised virgin purity, and in the 1700s white or silver wedding dresses were increasingly the choice of royal and aristocratic brides.

But the spectacular white wedding dates back no further than Victorian times. Throughout Europe until the 19th century people got married in their best clothes, whatever the colour, or in their national costume.

Bridal gown

When the first fashion plate of a white wedding dress appeared in the French publication *Journal des Dames* in 1813, the bridal gown was already becoming usual. The modest veil, which was worn in medieval times but had given way during the 18th century to a lacy cap or bonnet, also became an essential part of bridal wear in the early 19th century.

Decorative and symbolic flowers, first fashionable among the ancient Greeks and Romans, enjoyed a new vogue in Victorian times. Denoting chastity and fertility, they were worn by Middle Eastern Saracen brides, and medieval Crusaders brought the custom to Europe. Fertility was also symbolised in ancient Greece and Rome by the entourage of female bridesmaids accompanying the bride.

Throwing the bouquet to onlookers echoes the reversal of a medieval frolic in which wedding guests tossed stockings at the bride and groom. If the groom's stocking, thrown by a woman, landed on his head it signalled that she would soon marry.

The word 'honeymoon', from the honeyed wine that was consumed during wedding festivities, was probably coined in the 16th century. It evolved during the 19th century as a period of holiday and seclusion for the newlyweds.

Naming and christening
A child for God

TACITUS, THE 1ST-CENTURY ROMAN POLITICIAN AND HISTORIAN, RECOUNTS THAT SOME 2,000 YEARS AGO GERMANIC PEOPLES PUT THEIR BABIES' CRADLES IN TREE BRANCHES TO BE ROCKED BY THE BREEZE. THE WORD 'CRADLE' COMES FROM THE OLD HIGH GERMAN *KRATTO*, WHICH MEANS 'BASKET'. IN THE BIBLE, THE INFANT MOSES IS SAVED FROM THE EGYPTIANS BY BEING PLACED IN A BASKET AND LEFT TO FLOAT DOWN RIVER.

Adults, not babies, were the first people to be baptised as Christians. Initiated by the Church to symbolise spiritual rebirth, baptism marked the conversion of 'heathens' from paganism. 'Baptism', which derives from the Greek word meaning 'to immerse', reflects the common use of water in purification rituals before the advent of Christianity.

From about the 3rd century many newborn infants were baptised in a Christian ceremony that became linked to christening – receiving someone into the Christian faith. In the Middle Ages, when newborn babies had a tenuous hold on life, children were usually baptised on their day of birth, but as the risk of disease faded during the 1800s the period before the christening took place lengthened. Initially godparents were simply known as 'sponsors', from the Latin verb *spondere*, 'to promise', since they vouched for the sincerity of an adult undergoing baptism.

Food for growing infants
Breast, bottle and wet nurse

Throughout history, breastfeeding has evoked mixed feelings and not all women have welcomed this maternal experience. Wet nurses, hired by parents to suckle their infants, were common in Roman Britain and still much in demand in the 18th century. Queen Victoria refused to breastfeed, calling it 'making a cow of oneself'.

The Greeks and Romans favoured an urn-shaped terracotta cup as an alternative to breastfeeding. Hollowed-out cows' horns were common in Britain from the Middle Ages until the 18th century, while 16th-century portraits depict infants with feeders made from wood or pressed leather. Pewter feeders and pottery containers were both made from the 18th century.

The india-rubber teat, patented in the USA by Elijah Pratt in 1845, encouraged bottle-feeding, although there was still little understanding of the health risks involved, and sterilising bottles by immersing them in boiling water was not widespread until the late 1800s. The British boat-shaped Allenbury feeder of 1900, with openings at both ends, marked an important change. Not only was it easier to clean, it was the first feeder to include a teat and valve to regulate milk flow.

Feeding cups of the 17th and 18th centuries often contained caudle or posset, a custard of milk and eggs flavoured with wine and sugar. From the 1850s evaporated and condensed cow's milk was produced, followed by substitutes made from dried and powdered cow's milk in the 1880s. Pap boats for weaning babies on to solids were used in the 18th century. The pap was a mixture of breadcrumbs or flour cooked in water or milk.

Cradles and playpens
Keeping children safe

'Hush-a-bye baby, on the tree top, When the wind blows, the cradle will rock … '. The line from this favourite lullaby reflects the age-old awareness that gentle rocking gradually quietens a crying infant, as well as possibly recalling the ancient Germanic tradition of placing cradles in trees. Rocking cradles made from hollowed-out tree trunks are mentioned in English manuscripts of the 9th century, and by medieval times most cradles were either mounted on rockers or attached to fixed supports from which they swung from side to side.

Ingenious methods have long been employed to keep tiny children safely in one place. In his *A Regimen for Young Children* of 1473, Bartholomew Metlinger suggested constructing a 'little pen of leather'. A 17th-century 'baby minder' consisted of an oak post embedded in the ceiling and floor, with a wooden arm protruding at right angles. An adjustable waistband secured the infant to the extension, so that he or she could then 'run around'.

Baby clothes
Dressing the very young

Until the 1700s most infants spent their first months wrapped in swaddling bands. Swaddling was believed to encourage arms and legs to grow straight and had been practised since at least Roman times. But in the 19th century, an increasingly large layette was required: shirts, vests, caps, bodices, petticoats, gowns, shawls, bibs, socks and shoes. In the 20th century the list shortened, but babies' clothing reached its ultimate simplicity with the American 'Babygro' stretch suit, patented in 1959 by Walter Arzt, a father frustrated with dressing his child.

From the early 1900s, it became popular to dress boys and girls differently from birth. Choosing blue for boys' clothing was well established, recalling the ancient belief that blue granted protection against evil spirits. Girls, considered less important, needed no such aid; thus contrasting pink became the 'girls' colour', although this possibly recalled the 11th-century belief that red clothing warded off illness.

MANY HAPPY RETURNS

Birthdays were marked with feasts and gifts in Egyptian and Persian households of the 5th century bc. But celebrating children's birthdays only became popular during the 19th century ad, when societies began to stress the importance of family values. Birthdays were initially seen as a time for passing on lessons of morality and behaviour.

The ancient Greeks are thought to have honoured the birthday of Artemis, the goddess of the Moon and fertility, with candles placed on temple altars. If the candles were blown out in one go, good favour from the goddess was ensured. Today the accomplishment is said to make a silent wish come true.

May Day
The people's festival

In contrast to many pagan celebrations, Beltane and other European spring festivals held at the beginning of May were never incorporated into the Christian calendar. Only in 1978 was the nearest Monday to May 1 made a bank holiday in Britain.

Other spring celebrations in Britain included 'bringing in the May' – the bringing home of flowering hawthorn, first recorded in 1240. Maypoles have their origins in ancient European tree cults, but the first recorded example in Britain of festivities taking place around a Maypole dates to the mid 1300s – a Welsh poem describes the Maypole as a birch tree. The first known example of an English Maypole dance in which ribbons suspended from the pole were plaited and unplaited was in a performance of *Richard Plantaganet* by J T Haines in 1836.

The Puritans thought Maypoles to be phallic symbols and during the Commonwealth such 'stynkyng ydols' were banned. On the Restoration of the monarchy, Charles II commanded one of the tallest Maypoles to be put up on The Strand in London. At about 40m (130ft) tall, it provided a focal point for May celebrations for some 50 years.

April Fools' Day
Joker's paradise

Because some people will believe almost anything, they are the perfect victims for practical jokes on the morning of April 1. April Fools' Day, also known as All Fools' Day, has no clear origin, but is often said to derive from Hilaria, a Roman spring festival of jollity and dissipation. An alternative suggestion, which is based on the Bible, is that this was the day on which Noah released a dove from the ark, on what seemed to be a fool's errand, to find land on an Earth that appeared to be completely flooded.

One of the earliest known references to April Fooling in England comes in notes made in the 1680s by the antiquarian John Aubrey. A notorious practical joke involved a bogus 'Ceremony of Washing the White Lions'. In 1860 many Londoners were sent 'official' invitations to witness this ceremony at the Tower of London. They were to enter by the non-existent White Gate and many cab drivers repeatedly drove the length of Tower Hill looking for it. A similar trick is said to have been played at the Tower in 1698.

LA FOLIE FRANÇAISE

The fooling tradition may have come to Britain from France during the 16th century. There, April Fools' is known as *poisson d'avril* or 'April fish', *poisson* being a corruption of the French word for passion and a reference to the mocking and tormenting of Jesus by the Romans on the day of his Crucifixion, remembered around this time of year.

Easter celebrations
An annual rebirth

The long period of Lent ends on Maundy Thursday, which was named from the Latin *mandatum novum*, Christ's 'new commandment' that his disciples should love one another. In Britain the monarch gives Maundy money to several impoverished old people. Originally the monarch and the Church elders emulated Christ's humble act of washing his disciples' feet at the Last Supper, as well as making a charitable distribution of clothing, food and money to the poor. James II was the last monarch to perform this task in person – it passed to the royal almoner in the 1680s, before being dropped in 1754. By the beginning of Victoria's reign, the donation was of money only.

Easter Day is one of the Christian calendar's earliest festivals. Although it is not celebrated on a fixed date, Easter marks the start of spring. Its name comes from the fourth month of the pagan Anglo-Saxon year, *Eosturmonath* (April), said by the Venerable Bede to be named after the goddess Eostre, whose festival was celebrated then. In the 19th century Jacob Grimm, one of the Brothers Grimm, surmised that she was the goddess of dawn or spring. Others speculated that her emblem was the hare – the original Easter bunny.

The Easter egg is an ancient symbol of creation and rebirth. Murals on Etruscan tombs from the 6th century BC show it used in funerary rites, and the Greeks and Romans exchanged painted eggs at spring festivals. Zoroastrians, followers of a religion established in Persia around 600 BC, celebrate the new year on March 21 with gifts of painted eggs; new clothes are worn at this time, a custom echoed in the tradition of wearing something new on Easter Day.

By the 12th century Easter or Pace eggs, named from paschal eggs, were common to European Church rituals. They were boiled in dye and patterned with wax or flowers.

WHITE SUNDAY

Whitsun, also known as Pentecost, celebrates the coming of the Holy Spirit to Christ's disciples. The name derives from 'White Sunday', since on this day people wore white clothes for their baptism. From at least the 2nd century in England a week of festivities took place at Whitsuntide. One of the events was the Whitsun-ale, when villagers feasted and played games.

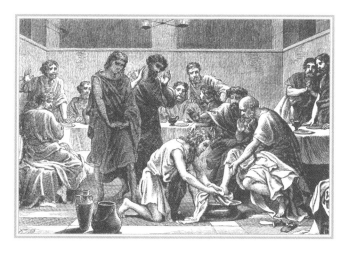

Harvest festivals
Celebrating the crop

An abundant crop has been cause for celebration since grain was first farmed and superstitions and rituals surrounding the harvest feature in most cultures. In ancient Egypt the harvest was overseen by Renenutet, the snake-headed goddess of fertility. Corn dollies may have originated in ancient Greece. In Britain the final sheaf harvested was believed to hold the corn's spirit and ears from this sheaf were plaited into a corn dolly, kept until the following spring. The last load of corn was also associated with the 'horkey', a feast to celebrate the harvest for those who had worked to bring it in.

THE USA GIVES THANKS

In the autumn of 1621, America's Pilgrim Fathers invited members of the Wampanoag tribe to a three-day feast in gratitude for their help during the previous year. By the mid 19th century a thanksgiving meal had become a yearly institution in New England. In 1863 Abraham Lincoln proclaimed that the last Thursday of November should be a day of thanksgiving, a tradition followed annually by subsequent presidents until 1941, when the fourth Thursday in November was established as a national holiday.

Increasing mechanisation during the 19th century eroded these traditions and today's harvest supper is linked with the Church's Harvest Festival. Before the Reformation in the 16th century, crops were blessed by priests, but subsequently the Church largely ignored the harvest until the mid 1840s, when the Reverend Robert Hawker instigated a thanksgiving festival at his church in Morwenstow, Cornwall. Within 30 years harvest thanksgivings were being celebrated in most parishes. The Reverend Henry Alford's harvest hymn 'Come, ye thankful people, come, raise the song of harvest-home' was first published in 1844.

All Saints and Halloween
In honour of the dead

For Celts and Anglo-Saxons the year ended when the herds were brought in from pasture at the end of October. The new year began in November and was marked by the festival of Samhain, a celebration in which purifying bonfires were lit. On the night before Samhain, souls of the departed could return temporarily to their hearths, and ghosts and demons were free to roam the Earth.

To counter the influence of this pagan festival, during the 9th century the Church instituted the feast of All Saints or All Hallows on November 1. Thereafter, October 31 became known as All Hallows Eve, or Halloween.

In 998 the abbot of Cluny in France established November 2 as All Souls' Day, when prayers are said for the departed, thus completing the link between Samhain and Christian festivals. In the 16th century, as they imposed Catholicism in Mexico, the colonising Spaniards took elements of local religions and incorporated them into their rituals. The Day of the Dead on November 1 remains one of the great celebrations.

In Britain, most Halloween traditions died out with the rise of Puritanism but games such as apple bobbing, in which apples floating in a bowl of water are caught in the mouth, are remnants of past rituals. The Halloween of today, in which children dress up as ghouls, ghosts and witches and light-heartedly demand 'treats' from neighbours under threat of a 'trick', has largely been reimported from the United States, where Irish immigrants introduced the custom in the 19th century. It is thought to stem from the idea that on Halloween all law is suspended. Dressing up in masks and costumes prevents people from being recognised by their own community.

The observance of saints' days, including All Saints, was banned by the English Protestants, but they soon created a new autumn festival in their place. In 1606 Guy Fawkes was hanged for his role in the Catholic plot to blow up the House of Lords the previous November. Parliament then passed an Act for the perpetual celebration of the failure of both this second attempt on the life of James I and the plan to destroy Parliament itself.

The Act ordered that everyone should attend church on the morning of November 5 as part of the Gunpowder Plot commemoration, which was soon popularly known as Guy Fawkes Day. By the 1630s the event was associated with bonfires and the burning of effigies. Initially these were of the pope or the Devil but from the 19th century they were often of Guy Fawkes himself.

New Year's Day
The calendar starts

Throughout the world the new year has regularly been linked to either the winter solstice or the vernal equinox. From 1155, when the English Crown first adopted the Roman practice of beginning the year in March, and up until the adoption of the Gregorian calendar in 1752, the English new year began on March 25, a date set by the medieval Church as the Feast of the Annunciation, or Lady Day.

Celebrations to mark this annual cycle were common in ancient Rome and continued over the centuries. In a masque by the English playwright Ben Jonson, which was presented to the court in 1616, a character appeared with an orange and a sprig of rosemary on his head to symbolise the season. In Scotland following the Reformation, the Church suppressed Christmas celebrations and New Year, or Hogmanay, became the main midwinter festival.

LIGHTING UP THE SKY

Fireworks were invented by the Chinese. From the 3rd century ad they celebrated the New Year by throwing bamboo 'firecrackers' onto flames. As the trapped air in the stalks grew hotter, they exploded with a bang. Around 1050 gunpowder, developed for military purposes, was added to the bamboo to create 'modern' fireworks.

Christmas cheer

'Well do Christian people call this holy day, on which
Our Lord was born, the day of the new Sun.'

St Ambrose, Bishop of Milan, AD 340–97

When the nights closed in and the shortest day of the year approached, people in ancient Persia and later in the far-flung Roman Empire paid homage to Mithras, the Persian god of light and guardian against evil. From 300 BC followers of Mithras lit fires around the time of the winter solstice to celebrate what the Romans called *Dies Natali Invicti Solis*, the birthday of the unconquered Sun. Worship of the god Mithras was one of the main rivals to early Christianity, but the fledgling Church was determined that the celebration of the Nativity should prevail and Christmas Day was firmly established as falling on December 25 by the 4th century AD. Nevertheless, the modern Christmas retains some elements of such pagan festivals as the Roman Saturnalia, which was celebrated around December 17.

Making merry

Named after Saturn, the Roman god of plenty, Saturnalia lasted for seven days and involved much feasting, drinking and merrymaking; homes were brightly lit and decorated with evergreens. Normal roles and customs were reversed: men dressed as women and women as men, while masters waited on servants, and otherwise forbidden pastimes such as gambling were allowed. A Lord of Misrule, elected from the servants of peasants, presided over festivities. This topsy-turvy world lives on today in the traditions of the Christmas pantomime.

The pre-Christian tribes of northern Europe also celebrated the Yule, a midwinter festival similarly marked – for those who had the means to do so – with indulgent eating and drinking and the exchange of gifts. Fires were lit, from which the tradition of burning a large yule log is descended. Homes were decorated for Yule with evergreens such as holly and ivy, which are symbols of renewal and everlasting life, and also with mistletoe.

Epiphany on January 6 celebrates the visit of the Magi to the infant Jesus and before the 4th century AD this day was the focus for Christian celebrations: in the Eastern Church it still remains so.

In 567 the Church declared that the 12 days between the Nativity and Epiphany were a sacred season. By the time of the Norman Conquest in 1066, this period was established in England as the year's main holiday, a time to rest and make merry. To keep the year's luck, Christmas decorations were taken down on January 6, Twelfth Night.

Festive traditions

Although Christmas celebrations are rooted in the past, they have evolved through the ages and continue to do so. Nativity plays, for instance, are 20th-century versions of medieval nativity pageants. Similarly, the medieval custom of carol singing in the streets was revived in the 19th century. Most modern carols date from that time, but one of the earliest, 'While Shepherds Watched Their Flocks by Night', was written as a hymn towards the end of the 17th century by the Irish poet and playwright Nahum Tate.

Today's Christmas is largely a Victorian invention embellished by commercial interests, but the first recorded Christmas tree was noted by a visitor to Strasbourg, then part of the Habsburg Empire, as early as 1605. The decorated fir is said to have been introduced into the British Christmas in the 1840s by Prince Albert. However, Princess Victoria, who was to become his wife and the future Queen, saw Christmas trees displayed at Kensington Palace in 1832, and in 1800 Queen Charlotte, the German-born wife of George III, had a tree put up for a party she held at Windsor on Christmas Day. Electric lights for the tree made their debut in the United States in 1882. They did so at the New York home of Edward H Johnson, an associate of Thomas Edison, who had created the first practical electric light bulb three years earlier.

Seasons greetings

In 1843 Sir Henry Cole initiated the Christmas card with a design by J C Horsley. Only about 1,000 copies were sold that year, but by 1880 the volume of cards was so great that the General Post Office had to ask people to post early for Christmas.

The Christmas cracker, said to be the brainchild of the English pastry cook and confectioner Tom Smith, dates from the mid 19th century. Smith's idea was to wrap a sweet and a printed riddle in twists of brightly coloured paper, but he was inspired to add the 'bang' after sitting in front of a crackling fire. He proudly advertised his crackers as 'combining art with amusement and fun with refinement'.

Christmas traditions continue to evolve. The monarch's message, first broadcast to the people of the United Kingdom and the Commonwealth in 1932, was originated by George V.

Signs and divination
The clairvoyant arts

For many years, the typical fortuneteller was a Gypsy, the nomadic descendant of a caste that once lived in India, where palmistry probably originated at least 3,000 years ago before spreading to China, Tibet, Persia, Mesopotamia and Egypt. In ancient Greece, where fortunetelling was highly popular and developed the essentials of its present form, Aristotle declared the hand to be 'the organ of organs, the active agent of the passive powers of the whole system'.

The early Church held that palmistry was the stuff of pagans and heathens, and denounced it as a dangerous form of sorcery and witchcraft. During the Renaissance, however, palmistry again found favour in educated circles as a legitimate science.

Foretelling the future in the pattern of tea leaves displayed in a drained cup has no recorded origins, but the practice was introduced into Europe from China with tea itself some 300 years ago and was well known in the 18th century. Coffee dregs were read in a similar way, and an announcement in a 1726 edition of the *Dublin Weekly Journal* mentions the arrival of 'the Famous Mrs Cherry, the only Gentle-woman truly Learned in that Occult Science of Tossing of Coffee Grounds'.

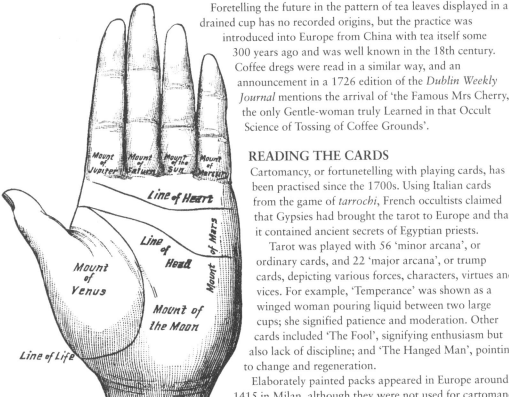

READING THE CARDS

Cartomancy, or fortunetelling with playing cards, has been practised since the 1700s. Using Italian cards from the game of *tarrochi*, French occultists claimed that Gypsies had brought the tarot to Europe and that it contained ancient secrets of Egyptian priests.

Tarot was played with 56 'minor arcana', or ordinary cards, and 22 'major arcana', or trump cards, depicting various forces, characters, virtues and vices. For example, 'Temperance' was shown as a winged woman pouring liquid between two large cups; she signified patience and moderation. Other cards included 'The Fool', signifying enthusiasm but also lack of discipline; and 'The Hanged Man', pointing to change and regeneration.

Elaborately painted packs appeared in Europe around 1415 in Milan, although they were not used for cartomancy at this time. Possibly inspired by variations of oriental packs brought to Venice by merchants, tarot cards may be named after the River Taro, a tributary of the River Po, which irrigates northern Italy.

Since 'reading the tarot' involves elements of myths and legends from many countries, including China, India and Egypt, its origins remain widely contested. Tarot may even derive from mystic cults that were suppressed by the medieval Church, but which re-emerged later as occult magic.

Astrology and horoscopes
Written in the stars

Thanks to the brainwave of a British newspaper in 1930, thousands of popular publications across the world now carry astrological forecasts, and almost everyone knows the 'star sign' under which they were born. In August of that year John Gordon, then the editor of the *Sunday Express*, commissioned the astrologer R.H. Naylor to cast a horoscope for Princess Margaret to celebrate her birth. As an afterthought, the piece also contained a few 'what the stars foretell' forecasts for people with birthdays in the following week. The reader response to these predictions was so enthusiastic that Naylor began a weekly column of horoscopes. This started a trend in British and American newspapers, and in 1936 the *New York Post* introduced the practice of running the horoscopes under the 12 signs of the zodiac.

Sumerian priests laid down the principles of astrology, a system of forecasting events by observing the movement of heavenly bodies, in the 3rd millennium BC. Some 2,000 years later, the Babylonians grouped stars into constellations from which zodiac signs take their names, and also divided the band representing the path of the Sun, Moon and principal planets into 12 equal parts.

PREDICTIONS AND DREAMS

✱ Before amniocentesis – the scientific analysis of fluid from a pregnant woman's uterus – was developed in the 1930s, superstitions offered the sole means of predicting the sex of an unborn child. Around 4000 BC in Egypt, two types of wheat were watered with the woman's urine. The child's sex was determined by which sprouted first.

✱ Around 2000 BC, the Egyptians devised guidelines for interpreting the images seen in dreams. A papyrus states: 'If a man sees himself in a dream looking at a snake – good, it signifies [many] provisions,' and 'If a man sees in a dream his bed on fire – bad, it signifies the rape of his wife.'

HEAVENLY FORECAST

In the 2nd century AD the Greek astronomer Ptolemy produced the first astrological textbook, *Tetrabiblos* (*Four Books*), stating that there were links between star signs and each of the four elements of earth, air, fire and water, and that these affected human destiny.

In subsequent centuries Roman, Arab and Renaissance scholars refined astrology, but its principles have endured. To obtain a true horoscope, the astrologer must know the date, time and place of someone's birth. The forecast is based on the position of the Sun and planets – and on the qualities traditionally ascribed to them – at that moment.

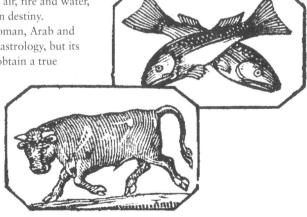

Superstitions, signs and omens

'He that hath understanding, let him count the number of the beast; for it is the number of a man: and his number is Six hundred and sixty and six.'

THE DEVIL'S NUMBER SPELLED OUT, REVELATION, CH. 13, V. 18

For the millions of people who cross their fingers for luck or to avoid retribution for telling a lie, superstition exerts a stronger grip than they might think. Like members of the earliest societies, who needed to control nature, they are seeking to ward off the random blows of fate. There is also a close link between superstition and religion.

Crossing the fingers for luck resembles the sign of the Cross enacted by early Christians. Touching wood to prevent the thwarting of a plan and holding a wooden crucifix when taking an oath, probably have similar origins.

Many plants and animals, because they are free of human control, have often been credited with supernatural powers. In North America and much of Europe the four-leaved clover, a rare form of the three-leaved variety, is thought to bring luck. Its properties were first recorded in the anonymous *Gospelles of Dystaues* of 1507, where it is said that 'he that findeth the trayfle [trefoil] with foure leues [leaves] … shall be ryche all his life'.

Hidden powers

Hawthorn is said to protect houses from being struck by lightning. It was long regarded as a protection against evil spirits believed to roam the Earth, especially on the eve of great festivals such as May Day. In the 17th century Francis Bacon noted the general belief that the plague could be detected by 'the smell of … May flowers'. This may explain the superstition that hawthorn flowers, or May blossom, invite bad luck if taken inside the house.

Like the hawthorn, the raven has associations both good and bad. Because this bird is a carrion feeder with jet-black plumage, it is often thought to foretell death and disaster. But the Spanish novelist Miguel de Cervantes relates in *Don Quixote*, published in 1605, that according to British tradition the legendary King Arthur had been transformed into a raven, and 'it cannot be proved from that time to this, that any Englishman has ever killed a raven'.

In ancient Greek mythology the peacock was the sacred animal of the jealous goddess Hera. After the death of Argus, the guardian with the hundred eyes, Hera had his eyes transferred to the bird's tail so that they could witness the adulterous liaisons of her husband Zeus. In medieval Europe the bird's

feathers were associated with a strong belief in the evil eye – the power of certain people or animals to bring bad luck to others simply by looking at them – and its feathers are still thought to attract misfortune if taken indoors.

Good fortune is assured, however, to those who greet the first day of the month by saying 'rabbits'. A 1979 survey revealed that about 5 million rabbits' foot talismans were sold in Europe each year. In ancient Rome hares were associated with fertility and lust, and uncaged at the spring festival of Floralia. Because the hare was once thought to reproduce without the male, in Christian art it sometimes symbolised the Virgin Birth.

Symbolic objects

The iron horseshoe, once vital in transport, agriculture and warfare, has evolved into a good-luck symbol. In 17th-century England and North America people hung up horseshoes as an antidote to witchcraft. It is even said that Admiral Nelson had one nailed to the mast of his flagship, HMS *Victory*.

Opening an umbrella indoors is said to attract ill fortune. The origins of the object – named after *ombrella*, Italian for 'little shade' – and the belief lie in China, where umbrellas were first made around 1000 BC. To avoid insulting the Sun, which the Chinese worshipped, they were only opened outdoors.

Sneezes have been blessed since the days of the ancient Greeks and Romans. This is probably due to a primitive belief that such explosions from inside the head were signs from the gods – for either good or evil.

Magical maths

Numbers have long been endowed with occult significance. The number three, now associated with the saying 'third time lucky' and with accidents being anticipated in threes, was considered a mystical number even before the Christian Trinity. The Anglo-Saxons believed in the powers of nine, being three times three, and used it to heal and to break spells.

Seven imparts mixed blessings. The idea of a 'seven-year itch' reflects the widespread belief that life runs in seven-year cycles. There were also seven wonders of the ancient world, and Christians identified seven deadly sins. The modern superstition that it is unlucky to seat 13 people at a table may reflect the story of the Last Supper, when Jesus ate the Passover meal with his 12 disciples before being betrayed by Judas.

Mausoleums and monuments
Commemorating the dead

The pyramids of Egypt, constructed from around 2600 BC to house the tombs of the pharaohs, are magnificent tributes to the dead. The great edifice that lent the mausoleum its name, however, was the tomb of Mausolos, ruler of Caria, at Halicarnassus (now Bodrum in Turkey), built in the 4th century BC. One of the Seven Wonders of the Ancient World, it was destroyed between the 11th and 15th centuries, probably by an earthquake.

Britain's earliest stone structures to contain the dead were communal burial sites such as the West Kennet Long Barrow in Wiltshire, which dates from around 3400 BC. At their simplest, memorials in British churchyards consisted of compact piles of earth marked by a wooden 'grave-rail', a board on two posts bearing the deceased's name.

Funerary monuments erected by the major roads leading to Roman cities featured portrait busts and inscriptions. Less eminent individuals were accorded more modest versions of these early headstones. Similarly, urban cemeteries built in Europe in the 19th century were laid out with large monuments on their main paths, less showy headstones behind and unmarked paupers' graves far from the eyes of the casual visitor.

The colours of death
Symbols of mourning

The ancients thought that death was contagious. Wearing black helped to make mourners invisible to malevolent spirits that might seek them out. The Greeks wore black, while for the Chinese and Japanese white is the traditional hue of mourning, possibly because it is the colour of undyed cloth and therefore indicates humility and sorrow. In Iran, however, people believe that light blue will ward off the 'evil eye'.

In Europe, black has been a mourning colour since the late Middle Ages. Wealthy 19th-century families used it extensively, with black festoons hung around the door, black coffins covered with black palls, and walls and beds draped in black cloth. Black mourning dress and veils were customary, often

embellished with jewellery made of jet. Servants wore black, while carriages were garlanded with it. And not all was for show: Edwardian girls had black ribbons threaded through the edging on their underwear for months after a family bereavement. Poorer people wore 'favours' – ribbons tied to their lapels or bound around their arms – or other black accessories such as shawls.

CHAPTER 3
FOOD AND DRINK

It is likely that many of the foods we know today were discovered by accident. Bread in its current form was probably first made when some unknown cook mixed beer with her flour instead of water. The earliest ales were also likely to have been the result of a serendipitous accident when someone noticed the pleasant after effects of drinking fermented gruel.

That British favourite, the potato, did not arrive on these shores until the end of the 16th century, brought from South America by Spanish and Portugeuese explorers via a circuitous route. The starchy root vegetable had been a staple crop for the Incas for thousands of years but it took a few centuries before it became a mainstay of British cuisine.
Tomatoes, peppers and chocolate are also the legacy of the great explorers although many of these imports were initially treated with mistrust by Europeans, who feared they were poisonous.

In the late 19th century, improvements in transport and food preservation revolutionised the quality and quantity of food available. H J Heinz began canning food in 1888 and Clarence Birdseye found inspiration for frozen food whilst observing the Inuit storing frozen fish in Labrador.

The story of farming
Tilling the soil

Farming began where wild wheat and barley grew abundantly – the Zagros mountains of Iraq, the Taurus ranges of Turkey and the valleys of Jordan. About 12,000 years ago, farmers learnt to till the soil with simple draw-hoes made of hooked branches of wood, their points hardened in the fire.

In the fertile plains of Mesopotamia cereal crops brought prosperity. Instead of just supporting those who tilled it, the land could feed many. At Sumer, in the southern part of the region, farmers developed a seed drill for planting seed in rows – thousands of years before the birth of Jethro Tull, the English agriculturist credited with inventing it in 1701. The Sumerian device made a groove in the soil, the seed trickling into it through a funnel and tube.

After inventing the plough, probably during the 4th millennium BC, the Sumerians were able to break up and sow vast new areas. These first 'scratch ploughs' consisted of a heavy stick, dragged by one person and guided through the soil by another, to scrape a shallow furrow. By 3000 BC a more efficient plough had been devised which could be pulled by two oxen.

But heavier, damp soils required more powerful ploughs. These were in use in China around 200 BC, but only appeared in north-eastern Europe in the 6th century AD. They had three working parts: the coulter, which cut vertically into the ground; the ploughshare, which cut horizontally through the ground at grassroots level; and the mouldboard behind the two blades, turning the sods neatly over. Pulled by a team of oxen, they enabled farmers to clear virgin land. As a result, food production and populations both increased sharply.

Centuries later, breakthroughs in farming equipment transformed the prairies of the American Midwest into the granary of the USA. The English inventor Robert Ransome devised an all-iron plough in 1808, but the steel plough with an all-in-one share and mouldboard was the creation of an Illinois blacksmith, John Deere, in 1837. In 1892 John Froehlich of Iowa built the first successful petrol-powered tractor.

REAPING A HARVEST

To cut their grain, early farmers used flint-toothed sickles made of wood or bone. During the 1st century AD farmers in Gaul devised a mechanical reaper, but the Romans preferred to use their slaves to gather the harvest. The reaper fell out of use and labourers persevered with sickles. The Reverend Patrick Bell of Scotland designed a mechanical reaper in 1826, but Cyrus McCormick, a Virginia farmer, built a more successful machine in 1840

For thousands of years people beat the grain from the ears using a flail – a club-shaped head attached to a short handle by a chain or cord. A threshing machine to replace the flail was designed by the Scottish millwright Andrew Meikle in 1787. It consisted of a rotating drum shod with metal beaters, turning inside a metal screen. The corn was fed between drum and screen. British agricultural workers opposed its introduction, destroying many machines during riots in 1830.

In 1836, J Hascall and Hiram Moore of Michigan, USA, patented a machine to harvest, thresh, clean and bag the grain. Their horse-drawn 'combination harvester' did meet with some success, but it was almost a century before the combine harvester was established. The All-Crop harvester, produced by the Allis Chalmers Company in 1935, was cheap to buy and could be pulled by a low-powered tractor. Self-propelled combines were developed in the 1940s. Together with grain dryers, combines removed much of the uncertainty from harvesting, eliminating the need to stack the harvested wheat in 'stooks' in the field to dry.

Growing wheat
The essential grain

All the 20,000 varieties of wheat grown today descend from primitive grasses that grew wild thousands of years ago. Among these first wild wheats was *Triticum monococcum* or einkorn, 'one grain', which originated in Turkey. Another, emmer wheat, spread from present-day Iraqi Kurdistan to Egypt.

These were the first cereals to be domesticated, emmer probably around 7000 BC. Emmer was more robust, but was neither very productive nor suitable for bread. Around 8000 years ago it interbred with another wild species to produce the high-yielding *Triticum aestivum*, or bread wheat, still in use today. Together with club wheat (*T. compactum*), a softer variety now used for cakes and pastries, it gradually replaced emmer in the Middle East and Europe, although in eastern Europe the growing season was too short for wheat production and rye became the major bread grain.

Wheat did not become a universal food until immigrants from the Crimea introduced the hardy Turkey Red strain – named for its place of origin – into North America in 1874. It adapted so well to the new climate that vast quantities were grown.

Flour was first made by hammering the grains between two stones. The result was coarse, gritty and wore down the teeth. Querns, used everywhere that wheat was grown, may have been invented in the Near East. The miller placed the grain in a lower, concave stone, grinding it with another round or cylindrical stone held in his hand. Around 1100 BC the large, flat upper stone of the quern was fitted with a handle that was rotated as the grain was fed between the upper and lower stones.

THROUGH THE MILL

✤ By the 2nd century BC milling had become a profitable business in Rome with donkeys doing the hard work of turning the millstones. When the Romans introduced water mills throughout the empire large-scale milling became possible for the first time.

✤ Wheat was the driving force of the Roman Empire and on two occasions Julius Caesar attempted to invade Britain for its wheat, which was already being cultivated in the south-east by 331 BC.

Cultivating rice
The sacred grain

The most important staple food for the majority of the world's population is not wheat but rice, the seeds of another grass. The first known rice gatherers were the Neolithic Hoabinhian people, who lived in southern China and north Vietnam from about 10,000 to 4000 BC.

Evidence of cultivated rice dates back to around 6000 BC: grains of that period were found in the remains of an ancient village at Hemudu in Zhejiang, China. By 4000 BC, cultivation of long-grain and short-grain rice had become well established in China and was just beginning in India, where rice and barley later became known as the 'two immortal sons of heaven'.

The word 'rice', derived from the Aramaic *ourouzza*, passed to European languages via the Greek *oruzon* and the Arabic *arozz*. The Arabs introduced rice into their Spanish territories from AD 711, but it did not spread elsewhere in Europe until 1468, when Spaniards planted arborio rice, the short-grain variety that became synonymous with risotto, at Pisa in Italy. Rice came to England in the Middle Ages, in the spice ships from the Orient, and was a precious ingredient, used mostly in puddings. The 19th-century colonial influence of Indian cooking and the expansion of tastes to embrace Eastern cuisine in the 20th century has made rice popular in Britain.

The first loaves
Our daily bread

THE CROISSANT IS REPUTED TO HAVE ORIGINATED IN BUDAPEST IN 1686. NIGHT-SHIFT BAKERS HEARD HOSTILE TURKS TUNNELLING INTO THE CITY IN PREPARATION FOR AN ATTACK, AND RAISED THE ALARM. THE BAKERS SAVED BUDAPEST FROM INVASION, AND THEY BAKED A CRESCENT-SHAPED ROLL TO REPRESENT THE SYMBOL ON THE VANQUISHED OTTOMAN'S FLAG. THE ROLL CAME TO PARIS IN 1770 WITH MARIE ANTOINETTE, WHEN SHE MARRIED THE FUTURE LOUIS XVI.

Bread has not always been the staff of life. From about 10,000 BC Neolithic Britons made hearthcakes by grinding together weed seeds and wild emmer wheat, mixing the meal with water and cooking it on hot stones. Mixtures of flour and water gradually evolved into flatbreads such as Mexican tortillas and Indian chapattis. The first loaf of yeast-raised bread was probably created accidentally by mixing flour with ale rather than water to make dough, or from yeast spores in the air settling on dough before baking, particularly if the bakery was also a brewery. In Gaul, Spain and Britain people developed liquid raising agents or leavens from beer, while in Greece and Italy, grape juice or white wine were used. Greek bakers took their skills to Rome, and by 30 BC there were some 300 bakeries in the city.

Among the many breads produced in Roman bakeries was a flat bread topped with pickled fish and onions – the earliest form of pizza. Toast may have its origins in an English medieval dish that was made by toasting white bread, which was then soaked in a white wine, reheated and served with almond milk.

By the 18th century, new fermentation methods using milk, salt and barm, a yeasty foam from brewing beer, was producing finer loaves. In 1840, the secretary of the Austrian embassy in Paris established France's first Viennese-style bakery, where fine wheat flour, milk and water produced the long loaves that became classic French bread.

THE SANDWICH GAMBLE

The sandwich takes its name from the notorious British gambler John Montagu, Earl of Sandwich. In 1762, at the Beef Steak Club in Covent Garden, London, Montagu spent 24 hours gambling without a break. He asked for meat and slices of bread to be brought to him so that he could eat without interrupting his game. The convenient result appealed to London's fashionable society.

The mystery of maize
From corn on the cob to popcorn

Maize originated in about 7200 BC in Central America. It does not exist in the wild, and exactly how it was first grown is not known. However, *Zea mays*, or maize, may have developed from crossbreeding a wild grass known as teosinte with other grasses, or it could have accidentally mutated from a plant that was already domesticated. In the 1940s, archaeologists working at Bat Cave, New Mexico, found 5,000-year-old corn cobs. These early cobs – not kernals – were just 1cm (½ in) long.

The first European contact with maize came in 1492. On arriving in Cuba Christopher Columbus received two gifts: tobacco and the grain known in Haitian as *maïs*. It was first eaten in Italy around 1650 as polenta, a dish previously made with wheat flour, which had sustained the Roman army.

Sweetcorn, one of the most popular varieties of maize, was grown by Native Americans from at least the early 17th century, but was not widely cultivated until after the American Civil War. The Native Americans also introduced popcorn – they believed that inside each kernel was imprisoned a tiny demon that made it pop when heated.

Hearty oats
Cheap and nutritious

Farmers first noticed wild oats as weeds growing in wheat fields. They were cultivated in Germany, Denmark and Switzerland in the 1st millennium BC, and reached Britain in the Iron Age. Oats thrived in such cold, wet, upland areas as Scotland. Long before the Romans arrived in Britain the grain became a staple, cooked as porridge.

Not everyone appreciated the value of the grain. The Roman writer Pliny despised the Germanic tribes for eating what he and his contemporaries regarded as animal fodder. In his *Dictionary of the English Language* of 1755 the English writer Dr Johnson defined oats as 'a grain … fed to horses, fine live-stock and to men in Scotland'. James Boswell, Johnson's friend and biographer, responded: 'Yes, and that is why in England you have such fine horses and in Scotland we have such fine people.'

Game animals
Meat from the wild

THE ROMANS INTRODUCED GAME PARKS TO BREED FALLOW DEER, WHICH HAD REACHED EUROPE FROM ANATOLIA. IN ENGLAND THE LEAST DESIRABLE PARTS OF A DEER, SUCH AS THE TONGUE AND OFFAL, WERE USED IN A DISH OF MIXED MEATS CALLED 'UMBLES' – HENCE THE EXPRESSION 'TO EAT HUMBLE PIE'.

Wild animals and birds, eaten raw, gave humans their first taste of meat. Two million years BC *Australopithicus* was an African scavenger supplementing a diet of roots, berries, nuts and termites by hunting small game or fishing. By 400,000 BC early humans may have been using fire to cook pieces of venison.

To hunter-gatherer societies, 'home' was where game such as woolly mammoth, long-horned buffalo and deer could be pursued. In Britain hunters relied on the red and roe deer, not only as a source of food, but also for their antlers, which could be made into tools. Elk, ox and pigs roamed the vast forests that covered the country, and ducks and geese were also plentiful.

Originally from Spain, the rabbit was introduced into Britain by the Romans, who prized it as a delicacy. They raised rabbits and hares in 'hare gardens', but these declined with the end of the Roman occupation.

Left to fend for themselves, rabbits proved easy prey for wild animals and died out. But when they were reintroduced from France in the 12th century there were fewer predators, and rabbits that escaped from warrens established a flourishing wild population. Only the young were known as 'rabbits'; older animals were 'coneys'. Settlers named Coney Island near New York after its abundant colonies of the creature.

Chicken and other poultry
Birds for the table

The wild red jungle fowl, the ancestor of our barnyard chicken, was probably domesticated in India around 2000 BC. Originally a sacred bird and provider of omens, the chicken's culinary potential was discovered by the Greeks in the 5th century BC.

Celts from Gaul probably brought the first chickens to Britain during the 1st century BC. Roman recipes were used, such as a stuffing for roast chicken which included ginger, shredded brains, eggs and pine nuts. The Romans also introduced the guinea fowl, but it disappeared from Europe in the Germanic invasions following the fall of Rome. It only returned in the 1530s, exported to France by merchants from the Guinea coast of Africa.

Around 1523 the turkey came to Europe from Central America. In Honduras, Spanish explorers bemused by the strange, dewlapped bird called it a 'guineafowl chickenpeacock'. The English confused its origins with those of the Turkish merchants who first brought it to northern Europe. While most other New World foods caught on slowly, by 1570 the turkey was already popular Christmas fare.

Prime beef
Meat on the hoof

The fierce horned auroch, hunted by prehistoric people with bows and arrows or spears, was the ancestor of today's placid cattle. The massive beast was the last major animal to be tamed as a source of food, power and leather, but could be found wild in parts of eastern Europe until the mid 1600s.

Cattle, which were domesticated in about 6000 BC in Turkey or Macedonia, were probably only brought to Britain by farmers from northern Gaul in the 4th millennium BC. Beef was the country's favourite meat from the time of the Roman occupation. The long working lives and poor winter diets of cattle meant that their flesh was rather dry until crops such as Dutch clover were introduced in the 17th century. At this time the Yeomen of the Guard were nicknamed 'beefeaters' – a term first used for well-fed servants.

HOME OF THE HAMBURGER

The medieval Tartar nomads of the Eurasian steppes tenderised slabs of beef by keeping them under their saddles. Before eating the meat raw, they added salt, pepper and onion juice – the original steak tartare. In Germany this evolved into grilled Hamburg steak, a dish that emigrants took to the USA. In 1836 'hamburger steak' was on the world's first printed menu, at Delmonico's in New York. No one knows who thought of serving hamburgers in buns, but the dish was probably introduced at such fairs as the St Louis World's Fair in 1904.

Pork, lamb and mutton
Rearing pigs and sheep for food

Because they need to be raised on those foodstuffs that were an important part of the human diet, such as acorns, nuts and cooked grains, pigs were tamed much later than sheep. Britons were hunting wild pigs by 5000 BC but the more docile creatures we know today were originally bred from domesticated boars in China and Iraq around 2900 BC, when Neolithic people began to have surpluses of food, enabling them to breed pigs for their succulent meat and valuable fat and skin. Their bristles, too, were made into brushes – it has been said that the only part of the pig not used is its squeak. Pigs were the meat of the peasantry, and even city dwellers kept them. From ancient Greece to 19th-century New York, pigs roamed the streets feeding on refuse.

To the first farmers, wild sheep were ideal for taming. As well as being a source of wool and meat, they provided fat used for cooking, medicinal salves and tallow for rushlights. Sheep were probably first domesticated in Iraq around 9000 BC, a process helped, no doubt, by their natural gregariousness.

Around 3500 BC – in the same period that sheep were introduced into Britain from Gaul – shepherds founded one of the world's oldest civilisations in Mesopotamia. The Sumerians had some 200 words for the creature, including fattened sheep, mountain sheep and fat-tailed sheep.

Middle Eastern methods of cooking lamb arrived in Europe during the Crusades. Shish kebab, Turkish for 'roast mutton on a skewer', reached the West from Turkey, where soldiers impaled chunks of lamb on their swords, roasting them over open fires.

Harvesting the waters

'Let the waters bring forth abundantly
the moving creatures that hath life ... '

GENESIS, CH. 1, V. 20

Before primitive peoples hunted or fished they gathered seafood such as oysters, mussels, whelks, crabs and lobsters from shores and shallows. The empty shells became the first coins, jewels and implements. Bare hands were the earliest fishing 'tackle' until clubs and spears proved to be more effective. During the Palaeolithic era, people fished with barbed spearheads carved from antlers and bound to wooden shafts.

One of humanity's first tools was the gorge – the forerunner of the fish-hook. Used in the Dordogne in France by 25,000 BC, it was a piece of wood, bone or stone that wedged in the gullet of a fish and was attached to a line – initially a vine tendril, later a strand of leather or animal sinew. In about 2000 BC fish-hooks were among the first tools to be made from metal.

Ancient angling

By attaching a line to a stick, early anglers invented the fishing rod. For thousands of years, fishing rods were less than 1m (3ft) long. Harpoons were used in France and Spain from about 12,000 BC and the more practical bow and arrow followed in Europe around 8000 BC. Early fishermen also used a type of trident, the leister, a spear made from two or three barbed prongs of bone attached to a stick. In Estonia the skeleton of a Mesolithic fish was discovered with a leister prong still stuck in its back.

Funnel-shaped basketwork fish traps, possibly the first baskets ever made, were also made in the Mesolithic period. Soon after 3000 BC the inhabitants of Oronsay and Oban in Scotland made baskets to take crabs and lobsters from deep waters, but also fished from coracles in inshore waters, using lines to catch sea bream, haddock and skate. The first fishing nets were

made from twisted plant fibre, hair or leather thongs. Inshore fishing with nets began around 8000 BC, but the Greeks were the first to make sturdy boats to catch fish at sea, around 7000 BC. During the 4th century BC they also made the first steps towards oyster farming. Noticing that young molluscs attached themselves to pottery shards in the water, the fishermen of Rhodes threw more pieces into the sea.

The Romans were passionate about oysters and made an industry of rearing them: the first recorded oyster farm was near Naples in the 1st century AD. Oysters became so popular that they were imported from Britain. In post-Roman Britain, seafish, which were sacred to the Roman goddess Venus, were customarily eaten on Friday, *Veneris dies*, the 'day of Venus'.

Fast fish

As the Christian Church introduced its own rites, fish retained its association with Friday, which became a fast day in commemoration of Good Friday. Fish could be eaten on fast days, when meat was forbidden, because unlike meat, it was clean and bloodless. During the 16th century the reasons behind fast days became more economic than religious – they helped to preserve scarce meat animals and encouraged the ship-building trade.

Parsley and fish were seen as a natural combination by the 12th century, when the writer Alexander Neckham recommended serving fish with a 'green sauce' of parsley, sage, costmary (a wild daisy), dittany (a type of oregano), thyme, garlic and pepper. Later in the Middle Ages green sauces were associated with newly salted white fish, known as 'green fish'. A medieval recipe for *moules marinière* involved stewing mussels in wine, pepper and minced onion.

In the 19th century frying was seen as a way of masking the smell of stale fish and extending its keeping time. The result became popular as a cheap, handy food for the working poor. The first reference to fried fish as a convenience food occurs in Charles Dickens's *Oliver Twist*, serialised in 1837–39, in which he describes a 'fried-fish warehouse'.

In the 1850s and 1860s, fried cod or flat fish was sold with a slice of white bread or a baked potato. Among the claimants to the title 'Oldest Fish and Chip Shop in the World' is Lee's in Manchester. It was established in 1863, but did not serve chips until after 1870, when they were introduced from France.

Drinking milk
From mare's milk to cow's

Mare's milk was the choice of the first milk drinkers, the nomads of central Asia: cows were not yet domesticated when herdsmen first learnt to take milk from sheep, goats, asses and horses, soon after 9000 BC.

In Neolithic Britain, where cattle could easily fend for themselves in the forests covering most of the country, cow's milk was commonly drunk, but after the countryside opened up, in the Bronze Age, more sheep and goats were kept and milked.

The taste of sour milk – the result of hot weather and unhygienic containers – would have been familiar. Ancient Europeans drank whey and soured milk, and the Roman writer Tacitus referred to early Germanic tribes living on wild fruit, game and 'solid milk', which had been left to ferment.

Concentrating milk by evaporating and sweetening it, was one way of improving its keeping qualities. A British patent was taken out in 1838, but in 1852 an American named Gail Borden decided to develop his own condensed milk after a rough transatlantic journey during which the ship's two cows became too seasick to be milked. Initially the sweetened drink was unpopular with New Yorkers, who were used to milk whitened with chalk and made 'creamy' by the addition of molasses, but Borden's canned version gradually caught on, especially with soldiers in the American Civil War.

During his investigations into the souring process of wine and beer, the French biologist Louis Pasteur discovered that microorganisms from the air also soured milk, and that they could be destroyed by heating. In 1860 he 'pasteurised' milk for the first time by heating it to a temperature of 125°C (257°F), thereby caramelising the sugar to produce a dark-coloured milk. Pasteur later developed a method using lower temperatures and longer heating times, but the 'unnaturalness' of the pasteurisation process delayed its commercial introduction in Britain until the 1890s.

THE STORY OF ICE CREAM

Before ice could be easily obtained, water ices and ice cream were luxuries. In the 1st millennium ad the Chinese used snow and ice to make fruit-flavoured drinks and desserts. They passed on their knowledge to the Persians and Arabs, who made syrups chilled with snow. These were called sharbats, from the Arabic shariba, 'to drink'. Water ices appeared in Paris, Naples, Florence and Spain in the 1660s. When ice cream was first served in England, at Windsor Castle in May 1671, Charles II was presented with 'One plate of white strawberries and one plate of Ice Cream'. The hand-cranked ice-cream machine was invented in 1843 in the USA, and by 1850 ice cream was being sold commercially in Britain.

Creating cheese
Milk transformed

At first cheese was simply the soft curds left behind after whey had been drained from sour milk before drinking. Cheese-making was revolutionised by the discovery of rennet, a digestive juice that is secreted in the stomachs of some mammals. Rennet curdles milk, producing enzymes that ripen cheese

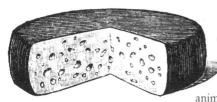

over a long period. It may have been accidentally discovered by Asian nomads who transported milk in bags made out of animal stomachs.

The earliest reference to cheese made with rennet in western Europe dates to the Romans. In Britain cheese has probably been made for 4,000 years. Julius Caesar found Britons preparing Cheshire cheese when he invaded in 54 BC. Cheddar, made from at least AD 1080, became famous in the 1600s, but was too expensive for most people. One 1727 recipe to make it at home may explain why: it recommends using curd made from milk with added cream and working in 1.5kg (3lb) butter.

Making butter
Milk churned

Butter may first have been made when milk was 'churned' by the movement of the containers used by central Asian nomads on their journeys. The name comes from the Greek *bouturon*, 'cow cheese', but although the Greeks knew about butter, they used it as a medicinal salve rather than as a foodstuff.

The Romans thought butter a food for barbarians such as the Celts, who probably introduced it into Britain in the pre-Roman Iron Age. As the Celts knew how to preserve meat and fish with salt, they would soon have realised that it would also improve the keeping qualities of their butter.

But the salt was rarely added with a light hand: records from 1305 show that 450g (1lb) of salt was being added to every 4.5kg (10lb) of butter. Before the butter could be eaten, the salt was removed by washing. The butter was then kneaded with water and the liquid pummelled out.

Eating eggs
From wild birds to domestic fowl

Prehistoric food gatherers probably collected wild birds' eggs to bake or to eat raw straight from the shell. Geese were being raised for their eggs in the Middle East by 2500 BC. Although chickens were known in Egypt by 1350 BC, in Persia, ancient Greece and pre-Roman Britain they were valued mostly as sacred birds or for their fighting abilities.

But elsewhere hens' eggs only became an important part of the human diet during Roman times. Perfecting Greek techniques for raising domestic fowl, the Romans produced hens that laid far more eggs than any wild bird could. As a result, eggs became widely used as cooking ingredients.

In China, people were eating eggs as early as 1400 BC and had developed a method of preservation that involved burying them in soil – hence 'thousand-year eggs'.

Garlic, leeks and onions
The poor man's spices

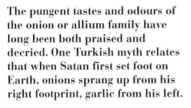

Onions probably originated in northern Asia and the Fertile Crescent and have been cultivated for some 5,000 years. In Mesopotamia and Egypt raw onions and bread were the everyday diet. With their white bulbs and long green stems they were more like spring onions.

From the late Middle Ages onions were a mainstay of European cooking. They were probably introduced by the Romans, who were keen onion-eaters. The abundance of wild onions in North America is recorded in the name of a city, Chicago, derived from the Native American word for 'onion odour'.

The pungent tastes and odours of the onion or allium family have long been both praised and decried. One Turkish myth relates that when Satan first set foot on Earth, onions sprang up from his right footprint, garlic from his left.

The rations of the Egyptian slaves who built the pyramids included garlic as well as chickpeas and onions, but the bulbs probably originate from central Asia. The Romans, who called garlic the 'stinking rose', brought it to Britain in AD 43.

Leeks are believed to derive from a Near Eastern variety of garlic. The first mention of their cultivation is in Mesopotamia in 2100 BC. They were adopted by the Welsh as their national symbol following a battle against the Saxons in AD 640. During the fighting, Welshmen wore leeks in their caps to distinguish themselves from the enemy.

Root vegetables
Buried treasures

Because edible roots were among the first of mankind's foods, their origins are often obscure. The radish has been cultivated for so long that its true origins are lost and its wild ancestor has long disappeared, but from China to the Mediterranean it was grown as a staple food plant. It was the Romans who brought radishes to Britain.

Domesticated in prehistoric times in its native Mediterranean region, beetroot was originally more like chard. The Greeks ate only the leaves, but turned the roots into a general tonic and blood cleanser. Beetroot as it is now known was first propagated in northern Europe during the 16th century.

The wild carrot of Afghanistan was small, thin and purple or yellow. Purple carrots may have been cultivated in Asia Minor in the 8th century BC. Neither the Greeks nor the Romans were impressed by the vegetable, although the Romans brought carrots to Britain. The Arabs took carrot seeds to Spain, from where the vegetable reached the Netherlands during the Spanish conquests of the 15th century. There the Dutch first produced a sweet-tasting orange variety by 1510. Around 1558 Flemish weavers escaping persecution by the Spanish brought orange carrots with them to England. At the court of James I their feathery tops were prized as decorations for hats.

High in the Andean mountains, where maize could not grow, the potato plant was first cultivated 5,000 years ago. The starchy tubers were a staple food for the Incas, and Spanish conquistadores took the potato home with them in 1539. Its journey to Britain was more circuitous but began in 1586 when the English admiral Sir Francis Drake returned home from the Caribbean. He collected his provisions, including some potatoes, from the Colombian port of Cartagena, before picking up settlers from Virginia. One of these gave potato tubers to the herbalist John Gerard.

Gerard assumed the plant was from North America, though in fact it did not reach that continent until 1719, via Ireland. He called it the Virginia potato to distinguish it from the sweet variety, known in Britain since 1565 when it was brought back from the Caribbean by the admiral Sir John Hawkins. The word 'potato' comes from *batata*, the Taino name for the sweet potato.

As it was the first cultivated crop not to be grown from seed, and because it ripened underground, the potato was feared as the devil's work. It was also thought tasteless and became established in Europe only after the French Revolution, when French chefs adopted foods with proletarian flavours.

> ## MYSTERIOUS MUSHROOMS
>
> Wild fungi must have been a useful source of food for hunter-gatherers. Mushrooms and truffles were delicacies in Mesopotamia by 1800 BC and in Rome. The Romans were mystified by their natural history. The ancients thought that, being rootless, fungi grew where lightning struck or where an 'evil ferment of the earth' remained.
>
> The Japanese were cultivating shiitake mushrooms some 2,000 years ago, but Olivier de Serres, agronomist to the French king Louis XIV, began mushroom cultivation in the West, using species still common in supermarkets today.

Cabbages and other greens
Leafy vegetables

Native to the coastal regions of northern Europe, including Britain, the curly leafed wild or sea cabbage was first eaten in prehistoric times. Cabbages, named from the Latin *caput*, 'head', were much appreciated by the Romans. A high demand for the vegetable initially made it too expensive for all but the wealthy. The emperor Claudius once called upon the Senate to vote on whether there was a better dish than cabbage and corned beef. The Senate dutifully decreed that there was not.

All other relatives of the cabbage are variations bred to favour certain characteristics – hence Mark Twain's description of the cauliflower as 'nothing but a cabbage with a college education'. Originally from Cyprus, the cauliflower was known to the Romans but reached Britain from France only in the 16th century.

Brussels sprouts, first described by the Dutch botanist Dodonaeus in 1554, have more mysterious origins: they may have been brought to the Low Countries by the Roman legions. The British did not discover Brussels sprouts until the 17th century, when they quickly became popular.

Appetising apples
The forbidden fruit

The Book of Genesis does not identify which fruit Eve tempted Adam to eat in the Garden of Eden, but medieval Old Testament scholars presumed it was the familiar apple of orchards everywhere. Thereafter it became known in Latin as *pomum*, 'the fruit of fruits'. Before being renamed, many new fruits and vegetables, such as lemons, tomatoes, potatoes and aubergines, were all called apples too.

Pips dropped by travellers helped to spread apples from their native south-eastern Europe and south-western Asia. Carbonised fruit found at Çatal Hüyük in Anatolia has been dated to 6500 BC, while dried apples were an important food for prehistoric Swiss lake dwellers.

Neolithic Britons picked wild crab apples, but sweeter, cultivated varieties were introduced by the Romans. One of the earliest named apples was the pearmain, first recorded about AD 1200. Large, hard apples used for cooking were called costards, and the men who sold them in the street were soon known as 'costermongers'. During the 16th century a sharp increase in apple imports so alarmed Henry VIII that he dispatched his chief fruiterer, Richard Harris, to France to learn grafting techniques. On his return, Harris founded the first of Kent's apple orchards at Tenham Manor.

Pears and plums
Orchard delights

Native to Asia Minor, pears have been cultivated for more than 3,000 years. The Greeks and the Romans increased the number of varieties, but two of our most common pears are more recent introductions. John Stair, a Berkshire schoolteacher, raised a new type of pear in 1769, but it was named after the man who distributed it, a Mr Williams. The origins of the comice are grander: it was first grown in the gardens of Prince Louis-Napoleon Bonaparte in 1849.

The common European plum began as a hybrid of the tiny sloe and other forms native to Asia Minor. Pliny, the Roman writer, described a 'great crowd of plums' in Rome's orchards, including a Syrian variety known as the 'plum of Damascus' or damson. The Romans introduced improved varieties into Britain, as well as cultivated cherries.

In 1725 Sir William Gage brought back a French plum named the *reine-claude* as a compliment to the wife of King Francis I. In English, it became known as the green Gage's plum, soon contracted to the greengage.

Exotic fruits
Tropical bounty

When, in 1493, Christopher Columbus landed on Guadaloupe he discovered pineapples being cultivated extensively, as they had been in Central and South America from the 1st millennium BC. Carib Indians, who called the fruit *anana*, meaning 'excellence' or 'fragrance', hung them on their huts as a sign of welcome, but also grew hedges of the spiky leaved plants to deter strangers.

The flavour and fragrance of the fruit 'astonished and delighted' the Spanish adventurers and those who received the *pinas de las Indias* that Columbus sent to Spain. A courtier to King Ferdinand described the reaction to the one fruit that survived the journey: 'In appearance, shape and colour, this scale-coated fruit resembles the pine cone; but in softness is the melon's equal; in flavour it surpasses all garden fruits. To it the king awards the palm.'

Wild banana fruits are hard and full of seeds, so the banana plant was first cultivated for its fibrous leaves, useful for thatching and for wrapping food for cooking, while the flowers and soft cores were eaten as vegetables. It is not known when the banana became edible and soft enough to be enjoyed as a fruit.

Grown in the Indus Valley before 3000 BC, the banana reached the Near East and Egypt by the 7th century AD, when the Koran identified it, not the apple, as the Biblical forbidden fruit. In 1482 Portuguese explorers encountered bananas growing along the coast of West Africa and adopted a version of their local name.

The first bunch of bananas seen in Britain came from Bermuda in 1633. Londoners marvelled at the sight of it ripening in a herbalist's shop. With the arrival of steamships in the 19th century, it became possible to import the perishable fruit from the West Indies.

The first British-grown pineapple was presented to Charles II in 1672. Thereafter pineapples carved in stone became an architectural motif, gracing the gates and roof corners of the homes of the wealthy as symbols of hospitality.

Berries and currants
The taste of summer

The warm, dry climate that Britain enjoyed after 2500 BC allowed its early inhabitants to pick naturally sweet wild strawberries and raspberries, although the former were hard to gather, being even smaller than today's wild raspberries.

Strawberry cultivation began in the 13th century, but the large modern fruit appeared only in 1819, the result of crossbreeding a small, sweet, scarlet fruit from Virginia with a pale Chilean variety tasting of pineapple.

Because the wild fruit is so full of flavour, raspberries were not cultivated in Britain until 1548. Gardeners also began to plant red and black currants, which they called raisins, thinking they were the fresh version of dried currants.

A TASTE OF THE ORIENT

✣ Apricots first grew wild in China and were cultivated as early as 2200 BC. The ancient Greeks called them the 'golden eggs of the Sun'. Peaches, which are also native to China, were known from the 5th century BC.

✣ The kiwi fruit was originally called the Chinese gooseberry, named after its place of origin. Farmers in New Zealand changed the name when they tried to introduce it into America in the 1950s, when anti-communist sentiments were popular.

Chocolate fancies
Confectionery favourite

The beans of the cacao tree were used to make drinking chocolate hundreds of years before Italian and French confectioners first produced eating chocolate – in Italy it was made into rolls and sliced. But it was not until 1819 that François-Louis Cailler of Switzerland had the idea of selling chocolate in blocks and making it on an industrial scale.

The earliest reference to eating chocolate in Britain appeared in an 1826 advertisement for Fry's Chocolate Lozenges, 'a pleasant and nutritious substitute for food' when travelling. Cadbury's marketed chocolate as confectionery from 1842, and in 1861 they introduced assortments, known as 'fancy boxes'. As Quakers who disapproved of alcohol and stimulants, the Cadburys and the Frys believed chocolate to be inoffensive and sustaining, although in fact it contains caffeine.

In 1875 a Nestlé worker together with a foreman at the factory of Daniel Peter, Cailler's son-in-law, had the innovative idea of adding condensed milk, thereby creating milk chocolate.

Golden honey
Nature's bounty

Flying from flower to flower gathering nectar, it takes 300 bees some three weeks to make about 450g (1lb) of honey. This wonderful source of sweetness and energy has been exploited by humans for many thousands of years, and in ancient times it symbolised all that was sweet in life. The Promised Land of the Old Testament was 'flowing with milk and honey'.

A Spanish cave painting provides the earliest evidence of honey-gathering. Dating to the late Palaeolithic era, it shows someone robbing a wild bees' nest on a cliff. The honeybee was probably semidomesticated in the Middle East in Neolithic times. Egyptian bas-reliefs of around 2600 BC illustrate bees being subdued by smoke – the first portrayal of beekeeping.

Wild bees only became plentiful in Britain once flowers began to bloom in the newly cleared pastures of the Neolithic period. Beekeepers of the late Bronze Age would carefully cut away sections of branches or hollow tree trunks where bees swarmed and take them home. Later, they made hives from bark or wickerwork, which they covered in honey and aromatic herbs in order to attract new swarms.

The story of sugar
The ultimate sweetener

During their invasion of India in 327 BC the soldiers of Alexander the Great discovered a 'solid honey not made by bees'. They were the first Europeans to taste sugar, which the Indians had been refining since 3000 BC. Some theories place the origins of sugar cane in the Bay of Bengal, others in the Solomon Islands of the South Pacific, but the Indians were the first to cultivate it.

From India, sugar cane spread to Indochina, Arabia and Persia, where it was being made into 'loaves' by the 5th century BC. Its high price meant that for centuries sugar was valued more as a medicine than as a foodstuff.

Sugar cane came to Europe around AD 800, during the Arab conquest of Spain, and was still used medicinally. In medieval England, delicate children were given violet and rose-flavoured sugars to encourage good health.

From the 11th century more regular supplies were brought by Arab merchants and returning Crusaders, who had 'beheld with astonishment and tasted with delight … the cane growing in the plains of Tripoli'. Sugar entered the kitchens of the rich in the 13th century. It was often kept under lock and key and used to season savoury foods. In 1380 the French king Charles V sprinkled sugar and cinnamon over a dish of toasted cheese.

> ## MOUTH-WATERING MOMENTS
>
> ❋ The inhabitants of Scandinavia, Germany and Britain chewed birch-bark 'tar' some 9,000 years ago. Modern chewing gum was born in the 1860s, when Thomas Adams, an American businessman, began selling chicle, a gum from the sapodilla tree of Mexico's Yucatan desert.
>
> ❋ Pastilles were named after the Italian confectioner to the Medici family, Giovanni Pastilla, who accompanied Marie de Medici when she married the French king Henry IV. The royal children called Pastilla's sweetmeats bonbons, literally 'good goods'.

COLONIAL COMMODITY

Although supplies of sugar increased as the European powers planted canes in their New World colonies, prices rose steadily after a tax was imposed in 1651. But the taste for sweetness was already well established. Noting the blackness of Elizabeth I's teeth, the 16th-century traveller Paul Hentzner thought it 'a defect the English seem subject to, from their too great use of sugar'.

By the 18th century even the poor had to have sugar, and they would buy just a few ounces a week. Sweetened meat and fish dishes became rare as sugar was increasingly added to tea and coffee, puddings, cakes, biscuits and fruit pies. But not until the tax was finally removed in 1874 did sugar become a common commodity.

A pinch of salt
The first seasoning

As farmers of the early Neolithic era began to eat less game than their hunting predecessors, they learnt to enhance the taste of their food: vegetables and cereals, unlike meat, contain little or no natural salt. At the same time, the invention of fire-resistant pots led to the discovery of a new cooking technique in boiling, which removes much of the salt in food. The white crystals also proved invaluable as a preservative.

Early communities obtained salt in many ways: from natural surface deposits, by boiling seawater, by burning seaside plants and getting salt from the ash, or by mining underground deposits.

In Europe the earliest known salt mines were established in the Austrian town of Hallstatt, near Salzburg – both names mean 'Salt Town' – during the 1st millennium BC. In Britain the deliberate production of salt was first practised in the early Iron Age by the Celts. Under the Romans, Droitwich, in the Midlands, became the site of a saltworks: 'wich' meaning 'a place where there is salt'.

The demand for salt quickly made it an important commodity and some of the world's earliest trade routes linked sources of salt to human settlements. The Via Salaria, 'Salt Road', one of the oldest roads in Italy, connects Rome with the port of Ostia and its saltworks.

The Romans gave salt rations, which they called a *salarium*, to soldiers and civil servants. Even when replaced by money it was still known as a *salarium*, hence the word 'salary' and the expression 'Not worth your salt'.

Pepper and spices
In search of flavour

The lucrative trade in pepper and spices began as early as 1450 BC, when Egypt imported cinnamon. From Malaysia and Indonesia, traders carried the spice across 4,500 miles (7,240km) of open sea in canoes to Madagascar, then up the coast of East Africa to the Red Sea.

The first and most important of the oriental spices to reach Europe was 'Indian pepper', known in Greece by 431 BC. In ancient and medieval times 450g (1lb) of peppercorns could literally be worth its weight in gold. The Romans introduced pepper to wealthy Britons, along with ginger, named from the Sanskrit *sringavera*, 'horn-root'.

Saffron, the yellow stamens of *Crocus sativa*, grew wild in Italy. The Romans left the work of harvesting it to the Greeks, who had been cultivating the flower since ancient times. Such refined tastes lapsed after the fall of Rome until the Arabs planted *zafaran* in Spain in the 8th century AD. Nutmeg and mace were the last major spices to reach Europe, probably in the 12th century.

Strategically positioned between the East and West, Venice grew rich from its monopoly of the spice trade. In the late 1400s other European nations set out to find new trade routes. In 1493 the Genoese explorer Christopher Columbus landed on Hispaniola (today Haiti and the Dominican Republic).

One local hot seasoning earned praise from Columbus, who described it as a 'better spice than our pepper'. The Spanish called it *pimiento*, the masculine form of *pimienta*, their name for 'black pepper'. In Mexico, where it had been valued since around 6000 BC, the Aztec word was *tchili* or *chili*.

The Portuguese took chilli peppers to India in 1525. Until then the Indian *kari*, 'sauce', was flavoured with cardamom, coriander seeds, turmeric and cumin, and made hot with black pepper. Dried chillies later became the basis of another hot sauce. Tabasco is a Native American word meaning 'land where the soil is humid', a good description of south Louisiana, where the seasoning was invented by the American Edmund McIlhenny in 1868.

FACT AND FICTION

✷ Arab traders concocted stories to conceal the source of spices. Huge birds, they said, collected cinnamon and used it to make nests on cliffs.

✷ The Chinese imported cloves in the 2nd century BC, when courtiers chewed them to sweeten their breath for imperial audiences. They called them 'birds' tongues'. 'Clove' derives from the French *clou*, 'nail', which the spice is thought to resemble.

✷ Vanilla comes from the seedpods of a climbing orchid native to Central America. It was used by the Aztecs to flavour chocolate.

Herbs for health
Ancient aromatics

When she caught the nymph Minthe with her husband Pluto, a jealous Prosperine turned her into an aromatic herb. As this Roman myth suggests, mint has been known since ancient times. It was much valued for medicinal purposes, particularly for stomach and digestive problems. The Romans often rubbed mint on their tables before a meal, perhaps to ward off indigestion.

Wreaths of wild olives were worn by successful Greek athletes at the Olympic Games while victors at the Pythian Games wore wreaths of bay leaves, from the Mediterranean tree *Laurus nobilis*. The victorious Roman generals were crowned with wreaths of laurel, and from this tradition came the expression 'to rest on one's laurels'. For banquets, the Greeks preferred wreaths of parsley – named from the Greek *petroselinon*, meaning 'rock parsley', which may have been one of the five kinds known to the Romans.

Roman settlers introduced thyme, sage, borage, chervil, dill, garden mint, fennel, parsley, rosemary and marjoram into Britain as garden plants. They also favoured coriander, perhaps the first Mediterranean herb to be imported to Britain, during the late Bronze Age.

Rosemary, a native of Turkey and southern Europe and known in Latin as *rosmarinus*, 'rose of the sea', was a symbol of love and death. Also originating in southern Europe were sage, first used as a herbal remedy, and thyme, named for its sweet aroma. The Greek word *thymos* comes from *thyos*, 'incense'.

THE NOBLE CONNECTIONS OF BASIL ARE REFLECTED IN ITS NAME, WHICH COMES FROM THE GREEK *BASILIKON*, 'KINGLY'. IT WAS A SACRED PLANT IN ITS NATIVE INDIA, AND THE EGYPTIANS COMBINED IT WITH MYRRH AND INCENSE IN THEIR OFFERINGS TO THE GODS. USED IN THE KITCHEN SINCE 400 BC, BASIL REACHED ENGLAND IN THE 16TH CENTURY.

Dried and salted
Making the most of the harvest

Cave-dwelling hunter-gatherers of the Pleistocene epoch hung mammoth and bison carcasses to dry in the wind, then took them inside for cool storage. Although ignorant of the cause, the ancients nevertheless prevented decay by salting, drying or smoking food.

Cereals, lentils and peas, the most ancient dried foods, were used in Syria, Iran and Palestine by 8000 BC; grapes and figs were being dried in ancient Palestine and Mesopotamia. And as the Old Testament relates, Joseph's foresight in storing grain during years of plenty was instrumental in his rise to power.

Salt was first used to preserve fish and game birds in the Nile Valley during the 3rd millennium BC. Britons began salting food in the early Iron Age, when the climate was becoming colder and wetter, making it difficult to preserve food by wind-drying. Salt fish became an essential part of the diet where eating meat was forbidden on certain days by the Christian calendar.

The Egyptians pickled vegetables and meat by salting them in spiced brine. Among Europe's first vegetable pickles was sauerkraut, made by layering shredded cabbage leaves with salt and leaving them to ferment. Although known to the Romans, sauerkraut was 'lost' after the fall of Rome and reintroduced from China in the 13th century.

Preserving meat
Curing, smoking and making sausages

Preserving pig meat by salting or curing it of its propensity for 'going off' was a speciality of the Germanic tribes of Westphalia some 2,000 years ago. Dry-salted and smoked, Westphalian ham from wild pigs, or *Bachen*, from which our rashers may get their name, was popular with the Romans, who also favoured hams from the acorn-fed pigs of Gaul. Wiltshire hams, flavoured with molasses and kept pink with saltpetre, were known by medieval times.

All meat was so valuable that waste, even of an animal's entrails, was anathema. Thus thrift explains the creation of the sausage. The ancient Babylonians and Greeks enjoyed sausages, and France and Italy have sausage-making traditions stretching back 2,000 years. Black puddings were a Gaulish speciality.

The type of sausage made was originally linked to climate, so that in southern Italy dry sausages were favoured, while moister varieties were developed further north. Names for sausages, such

PORTABLE PROVISIONS

✤ Hunter-gatherers sustained themselves on long journeys with dried salted meat similar to South American charqui, from the Peruvian *echarqui*, corrupted to 'jerky'.

✤ Potted meats or fish, covered and sealed with a substantial layer of spiced lard, butter or suet proved the ideal convenience food for Elizabethan sailors and travellers.

✤ H J (Henry James) Heinz began canning food in 1888, although his now famous baked beans were not test-marketed in Britain until 1905.

as frankfurter (Frankfurt am Main) and romano (Rome), were acquired from their city of origin, but salami is from *salare*, Italian for 'to salt'.

The Sumerians were the first people to combine drying and smoking flesh and fish. Later, red herrings became the mass-produced food of Britain during the Middle Ages. These whole, salted fish were dried and smoked until highly coloured, hard and so resistant to decay they could be eaten a year later. In fox hunting, the practice of 'drawing a red herring across the path' to destroy the scent led to the colloquial expression 'red herring' to describe a diversion.

The first kippers – flattened, salted herrings smoked or 'kippered' over an oak-wood fire – are said to have been produced in 1843 by a Northumberland fish curer. John Woodger was trying to develop for herrings the method used for smoking salmon since the 1300s.

Canned food
Extending shelf life

The Industrial Revolution brought a new victory over hunger. The enabling process was sterilisation, its executor a French confectioner named Nicolas Appert. As he boiled sugar for his sweetmeats, Appert tried to adapt the technique for perishable foods. In 1795 he arrived at the principle of canning. After much experimentation, Appert placed the food in loosely sealed jars, which he heated by boiling and then sealed tightly.

Appert's jars were soon superseded by canisters – shortened to 'cans' – of tin plate. These appeared in Britain at the beginning of the 19th century, and in 1811 the first canning factory was opened in Bermondsey, London.

Early cans were first filled with food and then sealed except for a small hole in the top. The contents were heated to boiling point before the hole was soldered. Then the cans were reheated, sometimes with fatal results: in 1852 one newspaper reported an occasion when 'steam was generated beyond the power of the canister to endure. As a natural consequence, the canister burst, the dead turkey sprang from his coffin of tin plate and killed the cook forthwith.' Once successfully sealed, these early cans had to be opened with a hammer and chisel.

FREEZING FOOD MAY HAVE ORIGINATED WHEN NEANDERTHAL MAN, SURVIVOR OF SUCCESSIVE GLACIATIONS, DISCOVERED THAT FOOD LEFT IN THE ICE STAYED FRESH. THE CHINESE ARE KNOWN TO HAVE STORED SNOW IN CELLARS BEFORE 1000 BC. ICE HOUSES WERE RARE IN BRITAIN UNTIL THE LATE 1600s, ALTHOUGH THE PRINCIPLE WAS KNOWN. THE ENGLISH PHILOSOPHER FRANCIS BACON IS SAID TO HAVE DIED OF A CHILL CAUGHT WHILE COLLECTING SNOW TO STUFF INSIDE A CHICKEN TO KEEP IT FRESH.

Ancient pulses
Beans, peas and lentils

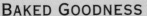

Nutritious and easily dried for winter storage, lentils, haricot beans, broad beans and chickpeas were some of the earliest vegetables to be cultivated. Lentils, probably native to the Middle East, were domesticated in around 8000 BC, while the various haricot beans native to the Americas were being grown by 7000 BC.

The Asian soya bean was first raised more than 4,000 years ago. According to Chinese legend, it was discovered by two bandits who survived being lost in the desert by eating an unknown plant. Descriptions of soya sauce have been found in Chinese tombs dating to around 160 BC.

In ancient Egypt, Greece and Rome, chickpeas, broad beans and lentils were daily fare for most people. Pulses were introduced into Britain from Europe in the pre-Roman Iron Age and used to make pottages. Peas were brought by the Romans.

A pulse known in Mexico as the *avacotl* was renamed the kidney bean by European herbalists after its shape and in the belief that it strengthened this organ. In 1597 the English herbalist John Gerard described boiled kidney beans and pods served with butter as being 'excellent delicate meat'. The runner bean, a native of South America, first came to Britain in the early 1600s, but was grown only as an ornamental plant until the 19th century.

BAKED GOODNESS

The first baked beans may have been the Egyptian dish ful medames, which in ancient times was made by burying broad beans overnight in hot ashes. In England, pork and beans was the staple fare of the Middle Ages. Settlers took the idea with them to North America, where they baked beans with pork and molasses. In 1875 the first canned beans were produced in the USA by Burnham and Morrill, but still with the original seasoned molasses sauce. Tomato sauce was introduced in 1891.

Nutritious nuts
Sustenance in a shell

For prehistoric peoples, the acorn was the most important nut in their diet, but the cave dwellers of northern Iraq in the Middle Palaeolithic era ate chestnuts, walnuts and pine nuts. Once woodlands were cleared for agriculture in the 5th millennium BC, the hazelnut took the acorn's place. Climate changes also meant that oaks and walnuts died out in Iraq and the Near East, and so almonds and pistachios are the only nuts mentioned in the Bible.

Among the early civilisations of the Indus Valley, the coconut was essential: the Sanskrit word for the coconut palm means the 'tree that furnishes all the necessities of life'. Its leaves were used for walls and roofs, its trunk for building and its shell as a vessel for eating and drinking. The hair of the mature nut made coir matting.

The ancient Greeks and Romans believed that walnuts, because of the kernel's resemblance to the human brain, cured headaches. In the 17th century, early settlers in the North American colony of Virginia tried to stave off starvation with walnuts, acorns, roots and berries.

The peanut or groundnut is actually not a nut but a pea-like legume. Originally from South America, it has been cultivated for at least 2,000 years, and peanut-shaped pottery jars have been found in Inca tombs. Portuguese or Spanish navigators introduced it into China in 1538 and then to Africa, where peanuts became a cheap food for slaves.

The coconut was known as the Indian nut until the late 1400s. Then the three dark spots on its base, like two eyes and a mouth, earned it the name *coco*, a Spanish and Portuguese word meaning 'grinning face'.

For centuries Peruvian Indians and Africans ground peanuts by hand to make a paste – the original peanut butter. In 1890 a St Louis doctor named Ambrose Straub invented a peanut mill. He introduced the nutritious spread at the World's Columbian Exposition in Chicago in 1893, but its early popularity was due to Dr J H Kellogg, the American inventor of cornflakes, who prescribed it for convalescents.

Essential oils
Extracts from seeds, nuts and fruit

For ancient peoples, vegetable oils were necessary for cooking, lighting and medicine. They were also used in religious ceremonies and for anointing the body. Following the advice of the Greek philosopher Democritus, who prescribed 'honey on the inside, olive on the outside', the Greeks and Romans oiled themselves after bathing, before and after exercise, and before and even during meals.

The oldest oil may be sesame, used in Egypt, Mesopotamia and Africa. The ancient Egyptians also extracted oil from radish seeds, while castor oil made a pungent choice for lamps. Almond oil was used in Asia Minor, coconut and soya in China and South-east Asia. Early American civilisations extracted it from peanuts, maize and sunflowers.

OLIVE ORIGINS
In Britain, as elsewhere in northern Europe, early farming communities used animal fat in their pottages and, when this was not available, linseed or the seeds of wild cabbage or native herbs. The Romans brought olive oil, but supplies dried up with the end of their occupation and butter became increasingly important. During the Tudor period Dutch immigrants cultivated rape extensively for its oil, and bright yellow fields became a familiar sight in south-east England.

It is not known who discovered how to extract oil from olives, or how to soak and brine the inedible fruit to make it palatable. What is known is that the olive tree was probably first cultivated in Syria and Palestine during the 4th millennium BC.

The olive seems to have reached Italy from Sicily in the 6th century BC and it was the Romans who spread it throughout Europe. With the screw press they also invented a method of making olive oil that has hardly changed. Virgin olive oil was produced from the initial pressing: second-quality and ordinary oils came from pressings of the pulp. Their oil did not keep and was often made just before using.

Margarine and other fats
Butter substitutes

France boasts more about its classic cuisine than about margarine, its other major contribution to modern eating. In 1867 Napoleon III launched a competition to find a substitute for butter that was longer-lasting and cheaper but just as nutritious. The only entrant was Hippolyte Mège-Mouriès, a chemist's assistant who had begun his experiments at the request of the French navy.

Starting from the premise that milk fat, the major ingredient of butter, was a type of body fat, Mège-Mouriès mixed animal organs with suet, a solid fat from around the kidneys, along with warm milk, minced pig's stomach and bicarbonate of soda. He called the resulting opalescent mixture margarine, derived from the Greek word *margarites*, meaning 'pearl'. By 1874 it was selling 100,000 tonnes a year.

In Britain, margarine was first sold in the 1870s as 'butterine', a product created by the Dutch butter merchants Jan and Anton Jurgens who had improved on Mège-Mouriès' process. Despite its name, it contained no butter.

The modern margarine industry was born when the process of hydrogenation was discovered by the French chemist Paul Sabatier in 1903. By using this technique, liquid oils could be solidified and thus transformed into a spreadable substance. Hydrogenation also turns polyunsaturated fats in liquids such as sunflower oil into the saturated fats often linked with heart disease. In the mid 1960s soft margarines high in polyunsaturates were first marketed as foods to be included in a healthy diet.

LOW FAT OPTIONS

Anxieties about the links between fats and ill health, have also led manufacturers to develop a range of semiartificial and artificial products designed to taste like fat but without the calories or cholesterol. The simplest are mixtures of starch and water turned into emulsions and used in such products as low-calorie mayonnaise. Emulsified proteins from milk and egg whites are used in Simplesse, which became the first fat substitute approved by the American Food and Drug Administration (FDA) in 1990.

Olestra is a synthetic compound of sugar and fatty acids that tastes like fat but passes unabsorbed through the body and has no calories. It was discovered in the early 1960s by biochemists at Procter & Gamble looking for an easily digestible fat for premature babies. Olestra is used in snacks such as crisps and biscuits – under its brand name Olean – although there are concerns about possible side effects.

IN PURSUIT OF HEALTH

✳ In 1935 the Japanese physician Minoru Shirota created a drink made from fermented milk, Yakult, that contained a lactic acid bacteria. His aim was to help people to maintain a healthy balance of bacteria in their intestines.

✳ Benecol, a full-fat margarine from Finland that reduces blood cholesterol levels, was introduced in 1996. Its key ingredient is sitostanol, a by-product of the wood-pulp industry, which inhibits the absorption of cholesterol.

✳ First marketed in 1985, Quorn is the protein produced by a mould that feeds on carbohydrates such as potatoes and rice.

Alternatives to sugar
Artificial sweeteners

To satisfy their craving for sweetness, the Romans created *sapa*, an artificial sweetener made by concentrating grape must in lead-lined pots. It was sweet but also poisonous – because of the pots it was stored in – and caused miscarriages, headaches and anaemia.

While investigating the reactions of certain coal-tar derivatives in 1879 at Baltimore's Johns Hopkins University, the American chemist Ira Remsen and a German student, Constantin Fahlberg, synthesised the compound orthobenzoyl sulphimide, and stumbled across a synthetic alternative to sugar. Having eaten some bread at his bench and found it sweet, Fahlberg realised that he must have had some of the substance on his fingers. Later analysis showed it was 300 times sweeter than sugar. Fahlberg filed a patent claim and obtained financial backing for the new product, which he named 'saccharine'.

One-tenth as sweet as saccharine but without its bitter aftertaste, cyclamate was discovered in 1937. Michael Sveda of the American chemical company DuPont was smoking in his laboratory and put his cigarette down on the edge of the bench. When he next put it to his lips he was struck by its intensely sweet taste. James Schlatter of the drug company G D Searle was searching for an anti-ulcer drug in 1965 when he realised that aspartame, a mixture of two amino acids – aspartic acid and phenylalanine – tasted sweet. Some 200 times sweeter than sugar, it was marketed in 1981 as NutraSweet.

> DISCOVERED IN 1879, SACCHARINE WAS SAID TO OFFER THE BENEFITS OF SUGAR WITHOUT THE DRAWBACKS. 'HERMESETAS DISSOLVE INSTANTLY, HAVE NO AFTER TASTE, AND GIVE YOU ALL THE SWEETNESS YOU WANT WITHOUT THE CALORIES YOU DON'T', CLAIMED ADVERTISEMENTS FOR A POPULAR SWEETENER IN THE 1960S.

Designer foods
Selection and engineering

Since people first kept animals and grew plants for food they have been crossbreeding their best livestock and their most abundant plants to create meatier animals and to develop crops with higher yields. In the 18th century agricultural innovators started to intensify the process, becoming even more selective about the animals they chose for breeding. A Leicestershire farmer's son, Robert Blackwell, was one of the pioneers, and created distinctive breeds such as Longhorn cattle, Leicestershire sheep and Large White pigs. Until the 1760s, British pigs retained the long-legged, slim form of the original wild boars; crossing them with plumper, short-legged Chinese pigs eventually led to the development of much larger, meatier animals.

When the Cambridge scientists Francis Crick and James Watson worked out the structure of DNA in 1953, they paved the way for genetic engineering in the 1980s. The first genetically engineered food to go on sale was the Flavr Savr tomato, in the USA in 1994. The fruit had a shelf life of 50 days or more. This was made possible by blocking the action of the gene that allows rotting.

The story of tea
Britain's favourite brew

One hot day more than 4,500 years ago, the Chinese emperor Chen-nung was boiling water to refresh himself. A few leaves blew off a nearby shrub and landed in the water, and Chen-nung named the resultant brew 'tay', or 'ch'a'. The legend is an appealing one but in fact no one knows who first drank tea. However, tea leaves do come from the evergreen shrub *Camellia sinensis*, which is native to the foothills of the Himalayas.

The Chinese first gathered tea leaves from the wild, and have cultivated the plant since at least AD 350. In time, taking tea took on social and cultural significance throughout the Far East and India, which culminated in an elaborate tea ceremony in Japan, popularised in the 15th century. Europeans first heard of tea in the 16th century from Dutch traders and Portuguese missionaries. Tea went on public sale for the first time in London around 1657. In 1660 the diarist Samuel Pepys sent for 'a cupp of tee (a China drink) of which I never had drank before'.

It was the court of Charles II that established tea as the fashionable drink of the elite. His bride, the Portuguese princess Catherine of Braganza, was a dedicated tea drinker, and after she arrived in England in 1662 the habit soon spread. The 17th-century trend for taking afternoon tea in tea gardens owed much to European adaptations of Oriental traditions.

The Chinese added heated milk to black, fermented teas, but not to green, unfermented ones. In 1665 tea was served with milk at a Dutch banquet in Guangzhou (Canton), and in 1671 the French brought the practice to Europe. As many teas were bitter-tasting, sweet-toothed Europeans also added sugar.

The ancient Chinese brewed up in unglazed red or brown stoneware pots and sent some of these to Europe with the first cargoes of tea. Dutch potters imitated the Chinese design, but soon gave their teapots more ornate and fanciful shapes, decorating them with coloured clays and glazes. The traditional, rounded British teapot is based on early Chinese pots.

AN ARABIAN LEGEND TELLS OF A GOATHERD MAKING A DRINK FROM BERRIES GROWING ON STRANGE SHRUBS AND THEN EXPERIENCING AMAZING LUCIDITY AND WAKEFULNESS. THE SHRUBS WERE SAID TO HAVE BEEN PLANTED BY DESCENDANTS OF THE QUEEN OF SHEBA, WHO HAD COME FROM KAFFA IN PRESENT-DAY ETHIOPIA, WHERE THE COFFEE PLANT *COFFEA ARABICA* ORIGINATES.

Invigorating coffee
A stimulating drink

The ancient Ethiopians made food balls of crushed coffee beans and fat to sustain them on long journeys. The beverage was developed in Yemen during the 10th century, and from the 15th century coffee drinking became widespread throughout the Islamic world, including Turkey.

The beans were first sold in Europe after a consignment reached France in 1644. However, as early as 1637 the diarist John Evelyn mentions a Greek drinking coffee in England. Britain's first coffee-house, The Angel, opened in Oxford in 1650, and coffee-houses appeared in London soon after.

Originally coffee of the best quality came from the Yemen's Mocha and Aden regions. However, in the early 17th century, plants were introduced from there into the Dutch colony of Java and into French colonies in the Indian Ocean, notably Madagascar. Coffee arrived in the New World in about 1723, when a French naval officer established a plant on the island of Martinique. A few years later, coffee was introduced into Brazil and burgeoned into a huge industry which was reliant on the slave trade of the 18th and 19th centuries.

In the ancient Arab world, milled coffee and water were heated several times to boiling point and the grounds left to settle before drinking. In Europe, coffee was made by a simple infusion until, in 18th-century France, a two-sectioned utensil was developed. The device separated the ground beans and hot water to produce a smoother drink, and came to be known as the cafetière. Count Rumford, the American-born Benjamin Thompson, devised the true coffee percolator – a coffee pot with a built-in metal sieve – in 1806.

> ## CUPS THAT CHEER
>
> ❖ Until the mid 19th century, tea was so valuable that it was kept in lockable containers, or caddies. The word caddy is derived from *kati*, a Malaysian measure of weight.
>
> ❖ Soluble instant coffee was the invention of the Japanese-American chemist Satori Kato in 1901.
>
> ❖ In 1903 Ludwig Roselius, a German importer, first made decaffeinated coffee, calling it Sanka, from the French *sans* (without) *caffeine*.
>
> ❖ John Lawson Johnston, a Scot, began production of his Fluid Beef in Canada in 1874. Back in London in 1886, he renamed it Bovril, from the Latin *bos*, 'ox' and Vril-ya, the 'life force', invented by the novelist Edward Bulwer-Lytton.

Drinking chocolate
The secret of cocoa

When the Spanish explorer Hernán Cortés came home from Mexico in the early 1500s, he brought back with him the Aztec sacred drink. Made from the roasted seeds of *Theobroma cacao* or the *cacahuaquchtl* tree, *xcoatl* or *xocolatl* was flavoured with a pungent blend of chilli, musk and honey. Although Cortés had grown to like this brew, the Spanish eventually added sugar, vanilla and cinnamon to make the bitter drink more acceptable to European palates.

The Spanish and Portuguese, who controlled the areas where the cocoa tree grew, guarded the secrets of *xcoatl* production jealously, but by 1606 cocoa paste was being exported to Italy and Flanders. It seems to have made its first appearance in London around 1657. By 1765 cocoa made with West Indian beans was being drunk in the American colony of Massachusetts.

Cocoa powder was developed by the Dutch chocolate maker Conrad J van Houten. In 1828 he patented a method for pressing cocoa butter from the roasted beans. This made possible the production of both instant cocoa and, with the addition of sugar and cocoa butter, chocolates.

Mineral water

The original health drink

The Romans appreciated the curative properties of mineral water, building spas from Bath in England to Tiberias in Israel. The idea of drinking, rather than bathing in, mineral water for its medicinal effects evolved more slowly, but was established by the Middle Ages.

For centuries scientists speculated about and tried to reproduce the therapeutic properties of naturally sparkling mineral water, especially its bubbles – known by chemists as 'fixed air' before finally being identified as carbon dioxide.

The pioneer of artificially carbonated water was the English clergyman and scientist Dr Joseph Priestley, who at first used carbon dioxide collected over vats of fermenting beer. In 1772 he demonstrated a carbonating apparatus in London. Three years later the English doctor John Mervin Nooth devised the 'gazogene', or soda siphon.

Artificially carbonated water was first produced for sale in Britain around 1781. But it was a jeweller from Geneva named Jacob Schweppe who, in 1790, established the first carbonated drinks company. After he moved to London in 1792, Schweppe's soda waters were sold by pharmacists for medicinal purposes and proved extremely popular. At the Great Exhibition of 1851, the company sold more than 1 million bottles.

Fizzy pop

From ginger beer to cola

People were already drinking soda water mixed with fruit syrups before the manufacture of prepared flavoured drinks began in the early 1800s. Ginger, popular since the late 18th century as the essential ingredient of ginger beer, was the first flavouring, added in 1820. Lemon followed in the 1830s. In 1850, ginger ale, developed by a Dr Cantrall in Ireland, revolutionised the soft-drinks market and was soon being sold in the United States.

Despite the different flavourings and sugar, carbonated drinks retained their medicinal connotations. In the USA, a pharmacist named John Pemberton based in Atlanta, Georgia, had had little success with such previous creations as Triplex Liver Pills and Globe of Flower Cough Syrup when he came up with a liquid cure for headaches and hangovers. The ingredients of his concoction included dried leaves from the coca shrub, from which cocaine is made (this was removed from the formula in 1905), extract of kola nuts and fruit syrup.

Advertised as an 'esteemed Brain Tonic and Intellectual Beverage', the syrup, now diluted with carbonated water, went on sale on a trial basis at

WHAT'S IN A NAME?

* Lithiated Lemon was the creation of Charles Griggs from Missouri, who introduced the lemon-lime drink in 1929. Four years later he renamed it 7-Up and sales increased sixfold.
* When the daughter of Newcastle chemist William Hunter fell ill with jaundice, he formulated a carbonated drink flavoured with orange and lemon oils and plenty of glucose. This seemed to do the trick and the drink was marketed in 1938 as Lucozade.
* Tizer – derived from 'appetiser' – was launched in 1924 by Fred Pickup in Manchester. He had begun in 1907 selling ginger beer, changing from stone to glass pint bottles in 1920.

Jacob's Pharmacy in 1886. One of Pemberton's partners, Frank Robinson, suggested the name Coca-Cola Syrup and Extract, while his flowing script provided the drink's trademark.

Pemberton sold 25 American gallons (94 litres) of syrup at $1 a gallon, but spent $73.96 on advertising. In 1887 he sold two-thirds of his ownership to an Atlanta pharmacist named Asa Candler for $1,200. After buying out Pemberton in 1891 for the sum of $2,300, Candler began his successful marketing of Coca-Cola as a drink for the young. Two years later he registered the Coca-Cola trademark, although the battle to register 'Coke' was not won until 1945. Coke's success inspired hundreds of other cola drinks, most famously Pepsi-Cola, first called Brad's Drink after Caleb Bradham, the pharmacist from North Carolina who created it in 1898 as a cure for dyspepsia.

Citrus drinks
Lemon, orange and lime

The Mongols were drinking sweetened lemon juice preserved with alcohol in 1299 – the first recorded people to drink lemonade. Northern Europeans had tasted their first citrus fruit only a century earlier, when the Crusaders came across the orange and lemon groves of Jaffa in 1191–92. Roman traders had probably brought citrus fruits back from India, but they were not cultivated in southern Europe until the Arabs, who had discovered them in Persia, began to propagate them around the eastern Mediterranean.

Towards the end of the 13th century, citrus fruits arrived in Britain. Seven oranges and 15 lemons, as well as 230 pomegranates, were bought in 1289 for Eleanor of Castile, wife of Edward I, from a Spanish ship at Portsmouth.

In Europe, the taste for lemonade developed during the 1600s in France. Carrying metal containers on their backs, the *limonadiers* sold their wares on the streets of Paris. The European café began with the shops set up by lemonade, coffee and chocolate sellers.

Initially, English lemonade was a potent drink; the first recipes included brandy and sack, a fortified white wine from Spain. Lime juice, added to the newly fashionable punch, was imported from the West Indies from the 1680s. By the early 1700s lemonade was preserved with sulphur and bottled. Although non-alcoholic lemonade had become common, wine mixed with lemonade remained popular.

Orange juice, however, was still rare. In 1669 the English diarist Samuel Pepys described his first taste: 'I drank a glass … at one draught, of the juice of oranges … they drink the juice as wine, with sugar, and it is a very fine drink; but, being new, I was doubtful whether it might not do me hurt.'

Tonic water was born after doctors advised British colonists in Africa and India to take quinine to protect them against malaria. Erasmus Bond patented an aerated quinine tonic water in 1858. From 1880 it was made by Schweppes.

From ale to beer
The art of brewing

Brewing only became possible once hunter-gatherers established organised, agricultural societies and began to cultivate cereals such as barley and wheat. Around 4000 BC the Sumerians learnt, probably accidentally, to make a crude ale, perhaps by noticing the effects of drinking gruel that had been left to ferment. The original ale was a weak brew made by soaking cakes of emmer wheat or barley in water and leaving them for about a day before draining off and then drinking the liquid.

Ale became more alcoholic once brewers malted the barley, a principle understood in Mesopotamia during the 3rd millennium BC. Germinating the grain, then drying and heating it so that the starch turned to sugar produced both a more powerful fermentation and a stronger flavour.

Fermentation itself was achieved by using the same vessels time after time. This allowed yeast – discovered by the French doctor Charles de La Tour in 1838 to be the microorganism responsible for the fermentation process – to build up in any cracks so that it could work on the next batch. Alternatively, the sediment from one brew was kept to start off the next, a basic version of the system used today.

The colour of the ale depended on whether the cakes were baked light or dark brown. Early brewers also balanced the sweetness of the malt by adding salt or perhaps mandrake, a plant with a leek-like flavour.

FROM HONEY TO HOPS

❖ The earliest known fermented alcoholic drink was mead, first made with honey long before the advent of farming.

❖ In Sumeria ale was also a currency and the name for it was *kash* – the origins of the modern common name for money.

❖ In 3000 BC a Sumerian poet praised the virtues of beer, claiming it left him in 'a blissful mood with joy in my heart and a happy liver.'

❖ Kreuger Cream Ale, produced by the Kreuger Brewing Co of New Jersey in 1935, was the first beer in a can. In 1962 tab-opening, all-aluminium cans were the brainchild of Ermal Cleon Fraze of Dayton, Ohio, after he forgot his can opener and had to bash open a can of beer on his car bumper.

GRAIN OVER GRAPE

In northern European communities, where the art of growing wine grapes was not known but where barley and wheat flourished, brewing probably developed independently. The earliest evidence of brewing in Britain goes back to the Iron Age and a Celtic brew known as *curmi*. In the 1st century BC, Saxons, Celts and peoples of Nordic and Germanic tribes were known for their ale. The Danes and Saxons who succeeded the Romans in Britain called their drink *öl*, which became the English 'ale'.

Beer was first brewed in Bavaria in the 9th century. It differed from ale by including hops, although in the 1500s these were also added to ale. Herbs had long been used in ale, and by medieval times hops were the plants most commonly used. Although native to Britain, hops were not used in local beers until the practice was introduced from Flanders in the 1400s. Bitter, well-hopped beer was preferred in southern England, sweeter ale in the north and Scotland. Both were drunk at all times of the day.

Most brewing was done during the winter. In Germany, beer that was kept cool in the summer with ice was called lager, from *lagern*, 'to store'. It only became commercially available when ice-making machinery was developed in the Industrial Revolution.

At the other end of the scale, porter, a dark, heavily hopped beer popular with market porters, was first brewed in London in the 1720s. It was exported to Ireland, but by the end of the century Irish versions were being imported to Britain, particularly porter, or stout, from the Dublin brewery established by Arthur Guinness in 1759.

Cider drinking
The apple beverage

Apple and pear trees were not seriously cultivated before the 5th century AD, when cider became a drink for the poor. The cider-apple tree originated in Biscay. It reached Normandy in the 11th or 12th century, after which cider successfully competed with beer, since using grain for alcohol could mean going without bread. Also known as 'apple wine', cider was brought to Britain by the Normans in the 1100s. Cider was made by crushing apples in a trough or a horse mill until John Worlidge, the English author of a treatise on cider, made a special press in 1676.

In the 1300s, government tasters tested an ale's strength by pouring some onto a bench and sitting on it. If their breeches stuck to the bench, the ale contained too much sugar.

From the 1620s English settlers in North America planted apple trees, partly because they knew they would need cider to drink, as at home, in place of disease-carrying water. In the 19th century Horace Greeley, the American journalist and reformer, complained that: 'In many a family of six or eight persons, a barrel tapped on Saturday barely lasted a full week.'

Public houses
Social drinking

Alehouses, taverns and inns were first established in Britain by the Romans. In the Middle Ages, many inns were set up for pilgrims and the influence of the Church was reflected in inn signs – necessary when most people were illiterate. The Lamb referred to Christ, the Crossed Keys to St Peter's insignia. During the Crusades the Saracen's Head appeared, while the title of England's oldest inn is claimed by Nottingham's Trip to Jerusalem.

Most publicans brewed their own ale, but from 1637 they were required to buy from a commercial brewer. In the 1700s brewers such as Bass, Courage and Tetley were founded and the 'tied house' was born.

A world of wine

'Wine is as good as life to a man, if it be drunk moderately:
what life is then to a man that is without wine? for it
was made to make men glad.'

ECCLESIASTICUS, CH. 31, V. 27

The pleasures of drinking wine began more than 7,000 years ago in the Zagros mountains of Iran, where the wild grape *Vitis vinifera* grew. There, around 5200 BC, as Neolithic people founded the first settlements, they discovered the potential of grape juice. Ripening grapes naturally gather a bloom of wild yeasts. Breaking the skin allows these to mix with the juice, so starting the fermentation process by which grape sugar turns to alcohol. To prevent the wine from spoiling, and to disguise any unpleasant flavours, the first winemakers added a resin called terebinth from the turpentine tree, *Pistacia atlantica*, which must have given the drink a powerful taste. By 2500 BC the Egyptians were making retsina-like wines, which they classified as 'good', 'good good', 'good good good' or 'sweet'. After harvesting, they threw the grapes into wooden vats and trod them with bare feet. The juice was then fermented in vats and filtered through linen. If not sold directly, the wine was stored in sealed amphorae labelled with details of the vineyard and the year.

According to Greek legend, Dionysus, the god of wine, fled from Mesopotamia because its inhabitants drank mostly beer. The Greeks, who had a flourishing wine trade, planted grapes in their colonies from the Black Sea to Spain. Greek settlers took vines to southern France in about 600 BC.

Pioneering vines

The Greeks also encouraged the production of wine in Italy, where it had probably been introduced by the Etruscans. When the Romans took over the old Greek vineyards they began to produce high-quality wines such as Falernian, a much prized white wine, creating an industry centred on individual villas. Poorer people quenched their thirst with diluted vinegar, from the Old French *vinaigre*, 'sour wine'. Roman soldiers offered this drink to Christ during his crucifixion.

From the late 2nd century BC the Romans encouraged vine cultivation in their provinces, including Gaul. Viticulture slowly spread further north, reaching the Rhône valley in the 1st century BC and then Bordeaux by the 1st

century AD. Italian wine was imported to south-east Britain before the Roman occupation – it was popular among the Celtic Belgae tribes who had invaded in 75 BC. Under the Romans it was also locally produced.

After the collapse of the Roman Empire, Europe's wine industry was saved by the Church as monks planted vines to provide wine for Mass. The Dissolution of the Monasteries in 1536 was a final blow to English wine. Production was in decline since Henry II married Eleanor of Aquitaine in 1152. Bordeaux had become part of the English Crown and French vineyards had expanded to meet the English taste for claret, from the French *clairet*, a pale red wine or a mixture of red and white.

Most wines were consumed within a year; fortified wines such as sack from Spain or the Canary Islands kept longer. Roman connoisseurs had thought Falernian at its best after 15 to 20 years, but the taste for mature wines died out after the fall of Rome.

Vintage wines were rediscovered after British glass-makers learnt, in the early 1600s, to produce glass bottles strong enough to store liquid. At the same time the use of cork to keep them airtight – known in ancient Egypt, Greece and Rome – became more widespread. The practice of binning wines on their side began in the 1730s, helping the ageing process by keeping the cork moist and stopping it from shrinking. It also led to a change in bottle style as the early onion shape gave way to cylindrical bottles in the 1760s.

Sparkling wines

These hardy new vessels played an important role in keeping the sparkle in champagne. At first the wines of the region were still. During the region's cold winters the fermentation process stopped, but it started again in the spring, giving off carbon dioxide. Winemakers initially felt that the gas was a problem, as fizzy wines were thought inferior.

But British drinkers enjoyed the drink's delicate flavour, to which they were introduced in the early 1660s. It was imported in barrels and bottled on its arrival in the spring. Sugar was often added, which boosted its natural effervescence. By 1664 the English writer Samuel Butler was referring to 'brisk [sparkling] Champaign'.

A Benedictine monk, Dom Pierre Pérignon, cellar master at the Abbey of Hautvilliers near Reims from 1668 to 1715, mastered the art of blending wines from different parts of the region and of producing fine white wine from black grapes. He also introduced the cork tied down with string, and strong bottles, but even so most of them exploded.

The French first began to take champagne seriously after 1715, when it was favoured at the court of the Duke of Orléans. Jacques Fourneaux established the first firm to produce champagne in Reims in 1734.

The beginnings of distillation
The water of life

The process of distillation was understood by the ancient Egyptians, Greeks and Romans who used it to make solvents and salves rather than alcohol.

In the 9th century the Arabs developed the art of distillation, making medicines, perfumes and elixirs by extracting the essences from fruits and flowers.

They also used the process to convert antimony, a metallic element, into an eye make-up called *al-kuhl*, meaning 'the powdered antimony'. This came into English as 'alcool', a word originally used to describe any fine powder or extract.

By the 1500s 'alcohol' had also come to mean the 'essence' of wine, but the word was not used solely for beverages until the 19th century. The names that were given to the first spirits reflected their fundamental importance to the many different people who enjoyed them: the Latin, Russian, Swedish, Gaelic and French for such spirits all translate as 'the water of life'.

SHAKEN, NOT STIRRED

When Xochitl, the daughter of an Aztec noble, served a drink of cactus juice to the emperor, he was so impressed that he married her and named the drink after his new wife. Another account traces the cocktail to Betsy Flanagan, an American barmaid who in 1776 decorated her bar with tail feathers from roasted poultry. When asked by a drunken patron for 'a glass of those cocktails', she served him a mixed drink garnished with a feather. Or mixed drinks may be named after the *coquetier*, 'egg cup', in which they were first made by a New Orleans apothecary named Antoine-Amadé Peychaud in 1795.

Brandy and cognac
The spirit of wine

In the West, distillation to isolate the essence of wine was first used by apothecaries in the Italian town of Salerno in the early 12th century. They heated the liquid in earthenware vessels, and the rising vapours were then cooled, condensed and collected. Because alcohol boils at a lower temperature than water, the result was almost pure alcohol. The earliest stills were known as pot stills, or alembics, from the Arabic word *al-anbiq*, 'the still'. Distillation derives from the latin *destillare*, 'to drop or trickle down'. As well as *aqua vitae*, 'water of life', this was known as *aqua ardens*, 'burning water'.

The apothecaries and doctors who practised the art of distillation used wine or the dregs of wine to make medicines. Monks distilled the juices of healing plants and evaporated these with water, or with wine, sugar and spices, to make the predecessors of liqueurs such as Chartreuse, first made by Carthusian monks in Grenoble, France, some 300 years ago.

During the Middle Ages *aqua ardens* became popular as a drink and by the 14th century it was being made by professional distillers. In Germany it was known as *Gebrandtwein*, 'burnt wine', which became the Dutch *brandewijn* and eventually the English 'brandy', a word originally used for all kinds of foreign spirits apart from gin.

Cognac gets its name from the region of western France where it is made. Dutch immigrants were the first to decide to distil the area's harsh wine. Once it was realised that the process of double distillation made a particularly fine brandy, the great cognac houses were founded, the first by a Jerseyman, Jean Martell, in 1715. He was followed by a winegrower named Rémy Martin in 1724, and by Richard Hennessy, an Irishman from County Cork, in 1765.

Whisky, gin and vodka
Grain-based spirits

The Celts called their 'water of life' *uisge beatha*, which evolved into *usquebaugh*, then *whiskeybaugh* and finally into 'whiskey' in Ireland and the USA, and 'whisky' in Scotland. Distillation may have reached Ireland from southern Spain, where it was introduced by the Arabs. By the 13th century, Irish religious houses were using barley to distil 'pot ale' – perhaps the earliest form of Irish whiskey.

Irish missionary monks travelling to Scotland took their knowledge of distilling with them. Although Scotch whisky had been made for some time before, the first written reference to it dates from 1494, when Friar John Cor is recorded as having bought 'eight bolls' of malt, enough to make 1,500 bottles.

Most whisky was made at home, using surplus barley which was malted then dried over a peat fire, giving the drink its smoky taste. Blended Scotch may have originated when barley farmers from the Borders took a cask or two of whisky to a local merchant who then combined them with other casks. John Walker, George Ballantine and the Chivas brothers, all began as licensed grocers making 'house blends'.

The Poles claim to have been making vodka since the 8th century, 400 years before it was recorded in Russia. In Russia the term *zhizennia voda*, 'water of life', evolved into *vodka* or *wodka*, 'little water'. The original recipe probably arrived by way of trade routes linking Russia and Poland with the Middle East. Most vodka is distilled from grains, but early makers used any available surpluses, including molasses, fruit and potatoes. The House of Smirnoff was founded in Moscow in 1812 and by 1886 it was the only vodka allowed on the imperial table.

Gin, the drink that earned the name of 'Mother's Ruin' in the 1700s from the drinking habits of wet nurses and mothers, was first thought of as a medicinal preparation. A type of gin was made in Amsterdam by 1575. However today's gin was the creation of a 17th-century Prussian-born physician named Franciscus Sylvius. He distilled a pure spirit alcohol from rye, and then redistilled it with crushed juniper berries. The name of the spirit, 'jenever' or 'geneva', comes from the Dutch word *jineverbes*, 'juniper'.

In the late 1500s, English soldiers fighting with the Dutch for the Protestant cause learned to appreciate a shot of 'Dutch courage'. They called the drink itself 'Hollands' or 'gin'. Dry gin came onto the British market in the 1870s, and was first made by Booth's in London. This drink was lighter than Dutch gin and proved popular mixed with soda water, bitters or tonic.

FROM 1655, THE ROYAL NAVY REPLACED BEER RATIONS WITH RUM, WHICH LASTED BETTER ON LONG VOYAGES AND DOUBLED AS AN ANAESTHETIC. RUM GAINED THE NEW NAME OF 'GROG' WHEN ADMIRAL EDWARD VERNON, WHO WAS NICKNAMED 'OLD GROG' AFTER HIS GROGRAM (GROSGRAIN) CLOAK, BEGAN DILUTING THE RATIONS IN 1740.

Exotic rum
Spirits from sugar

The West Indies gave the world rum, distilled from the sugar cane planted there by the Spanish from the late 15th century. The name may derive from the old Devonian word 'rumbullion', a slang word for 'uproar' or 'tumult'. White rum was made in Jamaica by 1825. From 1862 onwards it was produced by Emilio Bacardi, first in Cuba and from 1960 in Florida. The Cuba connection is still celebrated by two famous cocktails: the Cuba Libre, made from rum and Coca-Cola and created by a Florida bartender during the Spanish-American War of 1898, and the Daiquiri, consisting of rum, lime juice and sugar, which originated around the turn of the 20th century with American engineers working in Cuba's Daiquiri iron mines.

CHAPTER 4

FASHION AND BEAUTY

Western Europeans were wearing jewellery made from bone and ivory more than 40,000 years ago, many thousands of years before anyone thought to cover themselves with animal hides to keep out the weather. The jewellery would have had some religious or symbolic significance but it shows that man chose first to adorn his body rather than protect it from the elements.

The first clothing was made from natural materials – leaves, grass and animal skins – before people discovered how to weave cloth from wool, linen and cotton. Early garments were purely practical; keeping the wearer warm and relatively dry, or cool and out of the sun. After time, and as civilisations developed, different styles of clothing evolved. But the idea of fashion – that someone would choose to dress in a particular way because it was popular with other people – only really took hold in towns and cities in the mid 14th century. The growth of an international textile trade and the development of tailoring skills meant that clothes could be cut to improve or disguise the body's shape. At first, fashionable garments were the preserve of the elite but by the 19th century, style became a mark of individual freedom. And by the 20th century mass production meant that fashion was within the reach of all.

Fur, leather and felt
Nature's materials

The earliest fabrics were formed from large leaves and grasses pounded or joined together to make utilitarian clothes such as rain capes. In colder climates early humans learned to utilise gut thread and bone needles to sew animal skins into garments.

At first they used fish bones and thorns to hold the furs together at the shoulder, but by about 20,000 BC they were sewing garments similar to those worn by the Inuit today, with the fur on the inside. Fur appeared on the outside of clothes only from the Middle Ages.

Once people learnt to domesticate wild animals, the skins of sheep, goats, donkeys and cattle were commonly used for clothing. In Mesopotamia, the traditional costume of the Sumerians was a sheepskin kilt.

Clothes were made from rawhide during the Palaeolithic era, but they were stiff and perishable – in hot weather the hide dissolved into a glutinous mess. Possibly as early as 20,000 BC people discovered how to turn rawhide into leather. In Sunghir in Russia, burial mounds from that time have revealed skeletons clad in hats, shirts, trousers and moccasins made from leather or fur.

To create leather, the hair had to be removed from the hide and the skin softened with a mixture of oils, fat and brains. Inuit women wore their teeth down to the gums by chewing skins to make them supple. Then, after the fat layer had been scraped away, the inner layer was preserved by salting, smoking or tanning. Developed by the ancient Egyptians, tanning is a lengthy process that uses tannin-rich oak bark and galls – swellings on oak trees caused by insects or microorganisms – to soften the leather permanently.

Spinning and weaving
Warmth from wool

ACCORDING TO LEGEND, MERCHANTS LEADING CARAVANS IN THE MIDDLE EAST LINED THEIR SHOES WITH SOFT CAMEL HAIR, AND THE HEAT, SWEAT AND PRESSURE PRODUCED FELT. ONE OF THE EARLIEST MATERIALS KNOWN TO MAN, FELT DID NOT FRAY AND WAS IMPERMEABLE, MAKING IT USEFUL IN COLD, DAMP CLIMATES. IN BRONZE AGE GERMANY AND DENMARK PEOPLE WORE FELT CLOTHES, WHILE SIBERIAN NOBLES OF THE IRON AGE SPORTED FINE FELT STOCKINGS.

Before wool can be woven into cloth it must be drawn out into a continuous thread. Spinning probably began in the Zagros region of Mesopotamia, where sheep were first domesticated around 9000 BC. Wool fibres would have been rolled between the hands, but from about 7000 BC the spinster put the combed wool on a distaff, drew out the fibres and hooked them onto a spindle. The Sumerians were raising woolly white sheep by 3000 BC. Sheep were domesticated in Britain by 700 BC, but the Romans began wool weaving there only around AD 80. Yarns such as worsted, which originated in the East Anglian village of Worstead in the 1100s, helped to build the British wool and weaving industries.

The laborious process of spinning was speeded up by the invention of the spinning wheel, which turned the spindle mechanically. It arrived in Italy from India via the Middle East around 1300, reaching Britain 40 years later.

In 1733, English weaver John Kay invented the 'flying shuttle'. This shot automatically back and forth across the material, and enabled workers to weave varying widths of cloth so quickly that spinners could not supply yarn fast enough. In 1761 the Royal Society of Arts offered a prize for a machine that could spin six threads at a time and be operated by only one worker. The winning entry, invented in 1766 by Lancashire weaver James Hargreaves, spun eight threads at a time. It came to be known as the spinning jenny after a colloquial term for a machine.

Looms have existed in one form or another since 5000 BC. The most basic type of loom was strung around the waist of the weaver and attached to a tree, but this was soon followed by models with wooden frames to hold the threads in place. Thereafter looms became more complex, but they remained labour-intensive until the invention of the power loom by Leicestershire rector Edmund Cartwright in 1785.

Often using two or more colours, Scottish weavers made a type of twilled cloth with a diagonal weave from the 1400s. Around 1826 a London clerk making out an invoice for tweels, the Scottish spelling of 'twill', misread it and wrote the word 'tweed', a name that has stuck ever since.

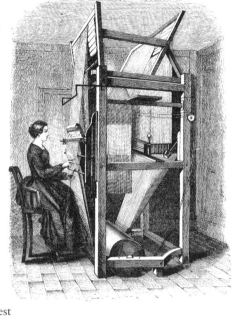

POPULAR PATTERNS

Tartans have existed for centuries in both India and South-east Asia. Originally a coarse material known in Old French as *tertaine*, tartan was made in Scotland from the late Middle Ages. Scottish clans began to adopt individual tartans as interest in a mythical past grew following the defeat of Bonnie Prince Charlie in 1746. But it was 19th-century British army regimental dress, together with the aristocracy, that popularised clan tartans.

The swirling paisley pattern originated in Kashmir, where for centuries colourful shawls were woven from Pashmina wool, later known as cashmere after the Kashmir goat from which it comes. These shawls were introduced into Britain in the 1760s and their immense popularity meant that copies were being made by the late 1700s. In the early 19th century, paisley shawls were manufactured in Norwich, Edinburgh and the town of Paisley, near Glasgow.

The original paisley design was based on the pine cone or the cashew, both of which are ancient symbols of fertility.

Cool linen
The first textile

Deep inside a cave in the Judaean desert, preserved by the dry air, the oldest examples of woven fabric yet discovered proved to be pieces of linen dating from around 6500 BC. Linen is made from *Linum usitatissimum*, the fibrous flax plant, which in Egypt was used some 2,000 years before cotton.

The ancient Egyptians collected wild flax plants from the banks of the River Nile, bound them into bundles, then dried them. The seeds were removed before the plants were soaked in water, which separated the fibres in the stems from the hard core, after which they could be beaten, washed, spun and woven. Cloth made from flax was strong, yet cool and light to wear. It could be wrapped or folded simply around the body. The Egyptians used sunlight to bleach their linen, which in its natural state is a dull grey-brown.

King cotton
The versatile fabric

On his return from India in 445 BC, the Greek historian Herodotus described 'a wool exceeding in beauty and goodness that of sheep'. Early civilisations prized cotton for its softness and whiteness. In the city of Mohenjo-Daro in the Indus Valley, people were weaving it into fabric from 3000 BC.

By 63 BC cotton was being exported to Europe from India and Egypt as a luxury material. It was introduced in Spain in the 10th century AD by the Moors, and 300 years later the first of Europe's cotton industries was underway in Barcelona, mainly producing sailcloth.

These fabrics still exist in the form of denim and jean. The fabric *serge de Nîmes*, contracted to denim in the 1500s, was made in Nîmes, France, where it was used as sailcloth. Jean, which unlike denim is dyed in solid colours or bleached, was first made in Genoa in Italy as a lining material for clothes and furnishings. It gained its present name in the 1560s, from the Genoese sailors who wore trousers made from the textile.

In England, cotton was known by the early 1200s, but no one had seen it growing. From 1612 the British East India Company began to import colourful Indian cottons such as calico, which was named after the Indian port of Calicut, and brightly painted chintz, from the Sanskrit *chitra*, 'variegated' or 'spotted'. At first cotton, particularly fine muslin from Al-Mawsil in modern-day Iraq, was prized for its rarity, but the increasing mechanisation of the cotton trade made it less expensive than wool by the early 19th century. European fashions were transformed. Women of all classes began to wear printed cotton dresses – the first time popular and fashionable tastes coincided.

Luxurious silk
China's best-kept secret

Around 2640 BC the mythical emperor Huang Ti asked his wife, Xi Lingshi, to discover what was eating the leaves of the mulberry trees in the palace garden. When a silkworm cocoon accidentally fell into some hot water, the empress pulled it out, only to discover that it was made from one long, delicate thread.

The origins of silk are shrouded in such legends. Certainly, however, it was the Chinese who discovered, in about 3000 BC, that the silkworm, which feeds on the leaves of the white mulberry tree, wraps itself up in a cocoon made from a single, continuous silk filament some 600–900m (2,000–3,000ft) long.

Silk first appeared in the eastern Mediterranean around 500 BC. East and West were linked by trade, and the routes along which the material travelled was known as the Silk Road. By 206 BC Chinese silk was being exported to Asia and the Middle East.

Industries for weaving and dyeing silk developed in Syria, Greece and Rome, but only small quantities of raw material were used. Most silk cloth from the East was unravelled and then reworked. Then, around AD 552, two monks smuggled mulberry seeds and silkworm eggs out of Persia in bamboo canes, thereby founding the European silk industry. England's own industry was established from 1685 by French weavers who settled in London.

The synthetic revolution
From rayon to Lycra

In 1664 Robert Hooke of the Royal Society first suggested the possibility of making an 'artificial glutinous composition much resembling' the substance produced by silkworms. If this solution could be formed into thin fibres, then they could be woven. Yet it was not until 1883 that the first man-made silk was created by the British inventor and physicist Sir Joseph Swan while searching for a filament for his electric lamp.

But Swan failed to exploit his new process commercially, leaving the Frenchman Comte Hilaire de Chardonnet to found the first factory for manufacturing artificial silk, or rayon, in 1890. His method was dangerous, however, and some of his early factories blew up. It was the British chemist C S Cross who, in 1892, patented the safer viscose process of making rayon, which is based on cellulose, the constituent of plant cells that gives fibres such as linen their strength. The earliest British maker of artificial-silk hosiery was the Wardle and Davenport company in 1912, but rayon was not widely used for clothes until the 1920s.

Nylon, the first all-synthetic fibre was developed by Wallace Carothers of the American chemical company Du Pont in the 1930s. Initially the new material was used to make toothbrush bristles. In the USA, women had to wait until 1940 for nylon stockings, while in Britain, delayed by wartime austerity, nylons first went on sale in 1946.

The principles used in nylon were soon adapted to create other synthetic fibres. Acrylic was sold in the USA from 1951 under the trade name Orlon. The British chemists James Dickson and Rex Whinfield discovered polyester in 1941, but it was declared secret by the Ministry of Supply until after the war. Patented as Terylene in Britain, it appeared in the form of ties in 1955.

The most successful of the new fibres invented in the 1950s were the ones that could stretch. Lycra, marketed by Du Pont in 1959, could increase up to eight times in length and still contract to its original size, making it ideal for underwear.

WASTE NOT, WANT NOT

❊ In China the bark of the paper-mulberry tree was used to make not just writing paper, but clothes and even armour. Surviving examples include a paper hat, belt and shoe dating from AD 418.

❊ Asbestos, which was known in China from the 3rd century AD, was called by the Chinese 'cloth washable in fire' and was even, on occasion, used to make suits.

The sewing machine
A servant in the house

The sewing machine brought affordable ready-made clothes within the reach of many, but at first few people wanted it. The English cabinetmaker Thomas Saint, who took out the first British patent for a sewing machine in 1790, probably never built one. Barthélemy Thimmoniers, a French tailor, created the prototype for the first commercially produced model in 1830, but his machines were destroyed by rioting tailors afraid for their jobs.

In 1834 Walter Hunt, an American inventor, created the lock stitch, the first true machine stitch. But his daughter persuaded him not to patent it because it would be 'injurious to ... handsewers'. Although all later machines used his lock stitch, Hunt failed to profit from it.

Nine years later, a Boston mechanic, Elias Howe, developed a machine after watching his wife's arm movements while sewing. He patented it in 1846, but was unable to find backing and in 1849 travelled to England. Howe sold the British rights for £250 to a corset manufacturer, but returned home destitute to find his patent infringed by the new manufacturers. These included Isaac Singer, who produced one of the first truly practical sewing machines in 1851. Howe successfully sued Singer, but the manufacturers subsequently decided to share ideas in one of the first patent 'pools'. It was Singer who forged ahead, however, introducing the home sewing machine in 1856.

In the 13th century a French tailor created the first pattern from thin wood, but the idea was discouraged by the tailors' guild until the 16th century. Help came for amateur sewers in the 19th century, when some women's magazines began offering their readers patterns for embroidery, lace and dressmaking. Packaged paper patterns, sized and with instructions, were the brainwave of Ebenezer Butterick, an American tailor. He created his first pattern, for a man's shirt, in 1863. Within a decade millions of his patterns were being sold across the USA and Britain, making fashion accessible to the middle classes.

THE SKILL OF BOBBIN LACEMAKERS LAY IN THEIR ABILITY TO PRODUCE EXQUISITELY DELICATE PIECES OF FABRIC, IDEAL FOR TRIMMING THE FINEST CLOTHING. ORIGINALLY, LACE WAS USED TO DECORATE UNDERGARMENTS, BUT BY THE 17TH CENTURY IT WAS REGARDED AS A LUXURY ITEM FOR TRIMMING BOTH MEN AND WOMEN'S CLOTHES AND WAS FASHIONED INTO CRAVATS, CUFFS, HEADDRESSES AND GOWNS.

Stitching techniques
The art of necessity

Quilting and patchwork were skills born of necessity and thrift. Both the Egyptians and the Chinese knew how to quilt by about 3400 BC, while the oldest known piece of patchwork in the world was put together around 980 BC in Egypt using the dyed and shaped hide of a gazelle.

The Chinese donned quilted clothes as protection against the cold, but quilting was also useful against a more deadly enemy: in 490 BC Persian soldiers at the Battle of Marathon had quilted armour. In the 11th century the Crusaders wore quilted clothing under their armour and brought the technique back to western Europe. By 1563 the trade in quilted bedding was established in London. In North America, English settlers of the 1600s recycled scraps of fabric to make patchwork quilts.

Simple oversewing to mend and patch skins and fabrics gradually developed into more complicated stitches and designs. Embroidery seems to have originated in China, Japan, Arabia and Mesopotamia before 1200 BC. In Britain the first reference to a specific embroiderer dates to AD 679, when Thomas of Ely described how St Etheldreda, the abbess of Ely, offered vestments worked in gold and jewels to St Cuthbert.

As it is known today, lacemaking is European in origin. Bobbin lace evolved from weaving in early 16th-century Flanders. It uses a pillow, a pattern and numerous threads, each of which is weighted by a bobbin. Needle lace, which began in Italy in the late 1400s, is created with a single thread and needle and is worked using thousands of buttonhole stitches to cover a web of threads.

> IN THE MIDDLE AGES PINS OF THE BEST QUALITY WERE MADE OF BRONZE, BUT IRON OR BRASS PINS WERE ALSO LUXURY ITEMS, HENCE THE TERM 'PIN MONEY', ORIGINALLY THE MONEY GIVEN TO A WOMAN BY HER FATHER OR HUSBAND TO SPEND ON SMALL ITEMS, SUCH AS PINS.

Pins and needles
The tools of the trade

Twenty-seven thousand years ago, people protected themselves against the cold with furs they had stitched together using bone needles pierced with a hole. Pins were made from fish bones, as well as from long thorns. Later, bone, horn and ivory were used, the needles having a round hole at one end or in the middle. From 4000 BC the Egyptians were fastening clothes with copper pins. Metal needles and pins, the wire bent over to form the head, were made in Europe during the Bronze Age.

To protect their delicate points, pins were being stored in cases from the 1300s; pincushions appeared in the mid 1500s. Pins continued to be made by hand until the 1820s, when Lemnel Wright, an American, developed a machine to do the job.

Steel needles were brought to Europe in the 14th century from the Middle East. The first European steel needles, which were produced in Germany in 1370, held the thread in a hook at one end. Metal needles with closed 'eyes' were being made in the Netherlands by the 15th century.

Thread, too, had been transformed by this time. Early sewers had used leather thongs, gut and grasses until technology provided linen, wool, silk and cotton yarns. In Britain thread of silk or linen, thought superior to cotton, was sold loose in hanks until spools were introduced in the mid 1700s. In 1844 the mercerisation process to strengthen cotton threads and give them a sheen was invented by John Mercer.

Thimbles were worn on fingers or thumbs, hence the Old English name *thymel*, 'thumb stall'. The first thimbles were conical and fashioned from leather, but later simple bands that left the fingertip exposed were also made – examples of these have been found in the Roman ruins of Herculaneum and Pompeii. The English were making dome-shaped thimbles from the early 1500s.

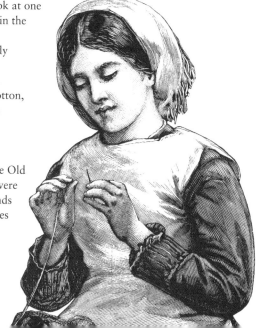

A suit of clothes

'In this month his Magestie and whole Court changed the
fashion of their clothes ... a close coat of cloth ... reached the
calf of the leg, and upon that a surcoat ... The breechs [were]
of the same colour as the vest or garment.'

RUGGE'S *DIURNAL*, OCTOBER 11, 1666

When, in the 14th century, the accounts for Edward III's
Great Wardrobe itemised several 'suits of clothes' –
probably the first recorded use of the expression – they
simply meant outfits of between three and six garments. From 1600
the masculine costume of a close-fitting jacket known as a doublet,
breeches and hose made up a form of suit, especially when all of
one colour.

Although the individual elements existed, the idea of a suit
of matching coat, waistcoat and breeches was not established
until the 1660s. In England it was Charles II who introduced
the suit, following a fashion set by the French king Louis
XIV. The London diarist Samuel Pepys described the new
mode on October 15, 1666: 'This day the King begins to
put on his vest ... being a long cassocke close to the body,
of black cloth, and pinked with white silke under it, and
a coat over it, and the legs ruffled with black riband like a
pigeon's leg.'

Under the coat

Buttoned the length of the front, the vest was the forerunner of
the waistcoat, although until the 1750s it often had sleeves.
Like the coat, which replaced the doublet, it was collarless,
knee-length and loose-fitting. The coat was left open to reveal
part of the rich fabric of the vest, but cheaper material was often used
for the back, hence the vest's early name of 'cheat'.

The 18th-century man of fashion had informal, dress and court
suits, but the difference between them was more in the fabric than
the cut, with court suits usually made of white, red, green or pink
silk or velvet. But in England and France from the 1760s there
was a move towards a simpler English style in dull shades of
brown, blue or black, based on the clothes worn by the
aristocracy for country pursuits.

By the second half of the century, the coat's full skirt had
been trimmed to create the tailcoat and morning coat.
Around 1770 the small coat collar appeared. As turned-down
coat collars grew higher, wigs and hairstyles grew shorter. By

1780 the vest had reached waist level. In fashionable circles the unsightly expanse of shirt that was now exposed by the shorter waistcoat was bridged by braces.

Trousers have been worn since ancient times: the Chinese dressed in trousers tied at the waist and often at the ankles to protect them against the cold, while Asian nomads wore similar garments for riding. In Persia they were traditional for women as well as for men, spreading to central Europe by 400 BC. Celtic peoples adopted trousers in the following century. English sailors and soldiers wore wide-legged trousers from the 1730s, but trousers only became fashionable after 1807.

By 1817 trousers were shoe-length and, by 1825, accepted as standard daywear. They were worn with a waistcoat and a full-skirted frock coat – first seen in the late 1700s – for the day, but with a tailcoat for the evening. The favourite patterns for trousers were strong plaids, stripes and checks. The loose, tubular cut was established in the 1860s. Trousers with front creases appeared in the 1880s and by 1913 had become common garb.

Jacket and tie

Not until the 1850s did the jacket become fashionable for casual wear. At first known as the sack coat, and later as the lounge jacket, its ancestor was the short suit jacket worn by boys and working men. In the age of mass production and the first ready-made suits, the jacket's simpler, more box-like structure was easier to make than the tailored coat.

By the 1860s the lounge jacket was being worn with matching trousers in plain materials, stripes or even checks. The lounge suit was adopted for informal daywear by the middle classes keen to present a respectable exterior, but businessmen kept their frock coats until the 1920s. Ties, worn since the 1830s, added a splash of colour. In the 1860s the Norfolk jacket, a version of the lounge jacket, appeared, worn with matching knickerbockers for outdoor pursuits.

Women's wear

For women, matching jackets and skirts appeared as holiday wear in the 1850s. By the 1890s tailor-made costumes were smart and practical everyday garments for young working women. The first trouser-suit to be worn by a woman was modelled by the French actress Sarah Bernhardt in Paris in 1876, but it was not until the late 1920s that women began to wear beach 'pyjamas'. Thereafter trousers were accepted for holidays, sports and country walking.

Skirt and dress
Fashion essentials

The earliest styled garments were draped around the body, accentuating its contours while marking the wearer's status and gender. In Egypt this form of dress developed before 1500 BC from the *schenti*, a belted loincloth that was

like a pleated, knee-length skirt. For women, the skirt finished under the breasts, secured by shoulder straps.

The styles adopted by the classical Greeks and Romans were looser. The chiton, worn in Greece from 700 BC, fell from the shoulders and waist in folds. The man's chiton was shorter, but otherwise there was little difference between the sexes. In Italy, the early Etruscans used ties to create close-fitting dresses, but contact with Greek civilisation led to the adoption of a draped chiton for Roman women and a flowing, semicircular toga for men.

In the 5th century AD the draped look was abandoned as the influence of northern Europe grew, spread by the Germanic tribes who invaded Gaul and raided Britain. From about 450 to 1300 the dress was a sewn, T-shaped garment belted at the waist. A flared skirt helped to accentuate the bust, waist and hips.

FIRST FASHIONS

Around 1350 the rise of competitive court societies in Italy and France introduced the idea of a fashion cycle. The dress now became the focus for the tailor's skill and the wearer's sophisticated taste. Lacing, buttoning and the cutting of fabric across the grain all contributed to a fitted look.

The floor-length gown became an exclusively feminine item in the early 16th century. From 1530 female costume was composed of the bodice and the skirt or 'kirtle'; the one-piece outfit remained as a long undershift. This form of dressing was worn for ceremonial occasions until the 18th century. For more private moments, the one-piece gown had re-emerged by 1700 as the loose, open-fronted 'mantua', which was often worn in the mornings.

In the 1800s advances in the technology of clothing production and use of different materials led to a wider variety of styles suitable for particular occasions. The width of the crinoline dictated the fashion for skirts, blouses and jackets in the 1860s, but from 1870 princess-line dresses were cut in one length.

By the turn of the 20th century, new sporting pursuits hastened the fashion for tailored skirts and blouses. The trend for sensible rather than symbolic forms of dress underpinned the simplification of female dress. From 1925 the short, shift-like dress encouraged the switch to mass production. It was left to designers such as Coco Chanel in the 1920s and Christian Dior in the 1950s to introduce stylistic features that reflected the mood and taste of the time.

Safety pins, buttons and zips
Fastening clothes

In the 2nd millennium BC in Greece, Italy or Sicily, the original safety pin was created by doubling a straight bronze pin and hooking one end into the other. The Romans wore safety pins as brooches, known as *fibulae*, to fasten their cloaks and robes. The safety pin then disappeared until it was reinvented in 1849 by the American inventor Walter Hunt. Finding himself $15 in debt, Hunt spent 3 hours one afternoon twisting pieces of wire to make a 'Dress Pin' with a coil spring and concealed point. A colleague paid him $400 for the rights, and the modern safety pin was created.

First seen by the 3rd millennium BC in the Indus Valley civilisation of Mohenjo-Daro, buttons were made from wood, bone, shells, stone, horn or pottery, and were more decorative than useful. The Romans, however, used them as fasteners by inserting them into loops sewn on the edge of a garment. Buttons were known in Britain from the 2nd millennium BC, but they were valued as fashion accessories. Girdles, sashes and clasps secured the loose clothes that were generally worn before fashions became more fitted during the 14th century, after the appearance of the buttonhole.

GRIPPING STUFF

While out hunting, Georges de Mestral, a Swiss aristocrat, was inspired to create an alternative to the zip by the burs that attached themselves to his clothes and to his dog. Examining the burs under a microscope, he realised they were covered in thousands of hooks and tried to reproduce the effect. The result, patented in 1956, was two strips, one with tiny loops to which the little hooks on the other clung. Although his fastener was made from nylon, de Mestral named it Velcro, from the French *velours*, 'velvet', and *croché*, 'hooked'.

BUTTONED UP
Hand-made from precious metals, glass or even gems, buttons were symbols of rank and wealth for the fashionable in the 14th and 15th centuries. Buttons became popular during the 18th century, when they began to be mass-produced from sheet metal. Even ankle boots were fastened by rows and rows of tiny buttons, thereby making the button hook an indispensable item. In 1893 a Chicago inventor named Whitcomb L Judson patented a 'clasp locker or unlocker' for boots and shoes, but his invention was as clumsy as its name.

Not until 1913 was the first practical 'hookless fastener' patented, by a Swedish-born American engineer named Gideon Sundback, who was employed by the Automatic Hook and Eye Company. The Talon Slide Fastener, as it was called, was used by US armed services in the First World War.

In 1923 the fastener was adopted by the American B F Goodrich Company for its brand of galoshes, originally called Mystik Boots. The enthusiastic president of the company renamed the boot the Zipper, and the name quickly caught on for the new fastening.

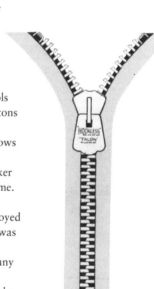

The shirt
Practical daywear

Evolving from the simple T-shaped undergarment of the Middle Ages, the linen shirt with embroidered front, cuffs and collar formed a distinctive and expensive part of the male wardrobe by 1530. By 1600, collars and cuffs had become such elaborate displays of lace and starch that they were separated from the body of the shirt and tied on.

This practice continued until the early 20th century, when stiffened cuffs and collars were attached by studs. The main part of the garment changed little, although the rise of the three-piece suit during the 18th century meant that less of the shirt was visible.

From about 1720 comfort, fit and respectability dictated the introduction of pleated sleeves, neck gussets and longer shirt tails. Completely open-fronted, buttoned shirts only became common after 1900, when stripes and colours provided some choice. By this time the blouse was acceptable as both day and evening wear for women.

During the 1920s American shirts with buttoned cuffs and attached collars announced a greater informality in male dress, with exotically patterned shirts designed for leisurewear. But in Britain the business shirt retained unbuttoned cuffs until the early 1960s.

Coats and mackintoshes
Against the elements

To keep out the cold, prehistoric people made cloaks out of animal skins held to their bodies with a leather thong. The basic design, used in ancient cultures and in the early centuries AD, was bell-shaped with a hole cut for the head – 'cloak' comes from the Old French *cloke*, 'bell'. For ordinary men and women cloaks were doubly useful, serving as clothing and bedcovers. More luxurious Egyptian capes were made of fine linen and hung from a decorative neckband.

The caftan, a sleeved garment worn crossed over in front and tied with a sash, originated in Asia in ancient times. It reached eastern Europe during the 13th century, when it was worn over trousers and fastened round the waist with a belt.

Early mackintoshes had some disadvantages. In 1839 the *Gentleman's Magazine of Fashion* remarked that 'a Macintosh is now become a troublesome thing ... on account of the offensive smell.' Until Charles Goodyear's vulcanisation process was applied to the cloth in 1843, it also melted in hot weather.

In western Europe the first true overcoat appeared in the late 16th century in the form of a calf-length coat with a fitted body and gathered skirt. Often blue, it was worn by apprentices, servants and those belonging to educational institutions such as Christ's Hospital in London. As a result, Christ's Hospital students were called bluecoat boys – and gentlemen avoided wearing the colour blue.

Only in the 1700s did the outdoor coat become a fashionable and practical alternative to the cloak and the cape. Initially men's coats were either top coats, styled identically to their indoor equivalents but made of a heavier fabric and considered suitable for walking, or capacious, heavy-duty greatcoats for travelling.

Coat styles proliferated during the 1800s. Around 1830, fashionable men adopted the knee-length waisted wool coat with a black velvet collar. The Raglan overcoat of 1857 – named after Lord Raglan, commander of British forces in the Crimean War – was distinguished by the cut of its sleeves, which instead of being stitched into a round armhole extended in one piece to the neckline.

During the Second World War the British navy developed a warm, short coat that fastened with wooden toggles. It was made from a woollen material known as Duffel after the Belgian town where it was first made commercially in the 19th century. After the war, large quantities of surplus coats were sold off cheaply and became a new 'uniform' for students.

KEEPING DRY

The search for waterproof clothing began when ancient peoples tied water-repellent leaves to a net base, or sewed strips of animal intestines together to create capes and head coverings. The Inuit made a type of waterproof hooded coat from strips of seal intestine or from seal or reindeer skin; this was known as a *parka* in the Aleutian Islands and an *anoraq* in Greenland. With the growing fashion for winter sports, 'anorak' was adopted by the Scandinavians in the 1920s for a lightweight, wind and rain-resistant sports jacket, while a heavier windcheater became known as a parka.

The first raincoat was created in 1747 in French Guiana by an engineer named François Fresnau, who waterproofed an old overcoat by smearing it with latex from rubber trees. Fox's Aquatic Gambroon Cloak was manufactured in London in 1821 using 'gambroon', a water-repellent twill, but it was Scottish chemist Charles Macintosh who first exploited the commercial potential of waterproof fabric. Macintosh sandwiched two pieces of cloth together with a solution of India rubber, but although waterproof, the resulting fabric was almost too stiff to tailor. Then a medical student named James Syme discovered a better method for dissolving rubber. In 1823 Macintosh patented the process using woollen cloth and began to make the waterproof coats that still bear his name.

In 1851 the Aquascutum company of London produced a chemically treated fabric that repelled water and could also be styled into the latest fashions. The name came from two Latin words meaning 'water shield'.

The trench coat had military origins. Manufactured in 1910 as the Tielocken by Thomas Burberry of London, it acquired its present name in the First World War, when a version that had rings attached to the belt for carrying grenades was made for army officers.

Gloves and scarves
Hands and heads

In ancient Egypt, Greece and Rome, gloves were worn for hunting, by soldiers, and for protecting the hands from hard work. Among the splendid treasures discovered in the tomb of Tutankhamun was a pair of white linen gloves dating back to the 14th century BC.

From the 6th century AD a pair of gloves was given to a bishop at his consecration to signify that his hands were clean for God's work. By the Middle Ages jewel-encrusted gloves were also seen as symbols of power. Despite being in disguise, Richard the Lionheart is said to have been taken prisoner on his way back from the Crusades because he could not bear to part with his costly gloves.

The leather gauntlets that were used for hawking and falconry were associated with the medieval knight, who would challenge another knight by striking him with his gauntlet or throwing it to the ground; hence the expression, 'throwing down the gauntlet'.

In 1834, Frenchman Xavier Jouvin invented a cutting die for gloves, based on detailed studies of hands in a dissecting room. The die, which could cut out six gloves at once, enabled gloves of a standard size and shape to be mass-produced for the first time. Gloves now became everyday wear for all and essential accessories that no fashionable woman, or man for that matter, would have been seen without. Whether it was long evening gloves or a delicate lace pair, there was something suitable for every social occasion.

COVERING UP

The origins of the headscarf and the handkerchief lie in the Anglo-Saxon headrail. Later known as the coverchief, from the Old French *couvrechef*, or 'head covering', it covered both the head and the shoulders, and hanged down the side of the face. After the upper classes abandoned the fashion in the 1400s, the kerchief, as it was then known, was still worn by peasant and working-class women until the 1800s.

Flowing scarves were an essential part of the feminine wardrobe around 1790 to 1830, reappearing in the 1880s with two American dancers who took Europe by storm. Dancing with the Folies Bergères in Paris, Loie Fuller used lengths of material stitched to batons, while Isadora Duncan incorporated yards of flowing drapes into her unique 'free-form' dancing. After the Russian Ballet also used scarves to sensuous effect from 1910, the possibilities of the scarf as a fashion accessory were quickly recognised.

FASHION FOR ALL

✻ The Reverend William Lee, who lived near Nottingham, developed the first knitting machine in 1589. Trying to attract a woman who was a keen knitter, he devised a frame to knit worsted stockings.

✻ The bibbed trousers known as dungarees take their name from a type of poor-quality Indian calico that came from the Dungri area of Bombay.

✻ In 1963 the invention of the plastisol transfer allowed people to choose their designs, reinforcing the T-shirt's status as a consumer classic.

Casual clothes
Fashions for every day

When designers such as Chanel made casual clothes fashionable in the 1920s and 1930s, knitwear became essential items in a woman's wardrobe. The origins of knitting remain obscure: it was possibly developed by Arabian nomads around 1000 BC, or in Egypt from the 7th century AD. The earliest examples date to the 1100s. By the 13th century knitting had reached Europe, where its popularity was reflected in the 'knitting Madonnas' painted in the late 1300s.

The people of the Channel Island of Jersey were known for knitting high-quality stockings by the 1580s. Jersey was the name for all knitted garments until, in the late 1600s, it was associated with the blue fisherman's jersey. Men adopted striped jerseys for winter sports in the second half of the 19th century. In the 1890s American college athletes coined the word 'sweater' to describe the garment's effect on their bodies.

WORK WEAR
Hoping to strike it rich in the Gold Rush, in 1853 a 24-year-old Bavarian named Levi Strauss left New York for San Francisco. But instead of prospecting, he established a wholesale business selling materials and other dry goods to miners.

In 1872 a Nevada tailor named Jacob Davis wrote to Strauss telling him of a method he'd developed of placing metal rivets to fasten pocket seams and the base of the fly. Davis asked if he would finance a patent with him. Strauss agreed and the patent was granted in 1873.

Made in denim with buttons for braces, a back pocket and a watch pocket, the 'waist overalls' proved so popular that two factories had to be built. When, around 1890, lot numbers were given to the products, the overalls with the copper rivets became lot '501'.

A 'short-sleeve white cotton undervest' was regulation wear in the US Navy from 1899. But American men only abandoned all-in-one underwear after coming across the French sleeveless cotton vest in the First World War. When Clark Gable appeared vestless in the 1934 film *It Happened One Night* undershirt sales fell. Then, in 1942, the US Navy made the 'T-type' vest regulation underwear. It was Marlon Brando's T-shirt-clad performance in *A Streetcar named Desire* that established the T-shirt as fashionable outerwear.

Evening wear
Dressing up

An evening dress code was established by the Englishman George 'Beau' Brummell. He perfected his sober style between 1800 and 1817, introducing the tailored black coat and black pantaloons as fashionable dress for evening engagements. Until then 'full dress' for a man had simply meant a more elaborate version of the suit he wore during the day.

This change coincided with a trend in the upper echelons of society for dividing the day – and fashions – into neat sections based on mealtimes. The concept of evening dress was in place by the 1820s, and in 1838 the Parisian paper *Le Dandy* noted that 'the black English suit with silk buttons is always required' for important occasions.

By the 1850s men's evening wear had become a uniform of black tailcoat, waistcoat, white shirt, white cravat and a pair of black trousers with a braid stripe sewn down the outside seam. The waistcoat was either black or white until the end of the 19th century, after which white was established as *de rigueur*.

JACKET REQUIRED

The advent of the dinner jacket was greeted with shock by 19th-century society commentators. Its place of origin is claimed as Monte Carlo, London and New York, where in 1886 the millionaire Griswold Lorillard wore this kind of jacket to a ball at the Tuxedo Park Country Club. In the USA it is still a 'tuxedo'.

Derived from the lounge jacket in the 1880s, the early 'dress lounge' – the term 'dinner jacket' was coined only in 1898 – was worn with a black waistcoat and a black tie. By about 1914 it was fastened with one button, and by the 1920s it was double-breasted.

But full evening dress was expensive to buy. In 1897 Charles Pond, an impoverished entertainer, asked to borrow a dress suit from Moss Bros, a secondhand clothing shop in London established by Moses Moss in 1860. Moss's son Alfred lent Pond a dress suit without charge several times before the men agreed on a fee, thereby establishing the earliest dress-wear hire firm.

The first designer whose name was as familiar as her work was influential was a Frenchwoman named Rose Bertin. In the 1770s she was dressmaker and milliner to the French queen Marie Antoinette, but died in poverty in 1812, her fashions out of date.

Until the mid 1800s fashionable women visited dressmakers who would custom-make outfits for each client in their workshops. The idea of the luxury haute-couture salon, where a designer showed his creations, emerged with the

establishment of the first fashion house in Paris. But although France was the acknowledged arbiter of fashion from the mid 17th century, the first haute couturier was an Englishman.

Born in Lincolnshire in 1825, Charles Frederick Worth was apprenticed at the age of 12 to a firm of drapers. Eight years later he left for Paris to become a designer, setting up the House of Worth in 1858. Although he took credit for the demise of the crinoline in 1870, his skill was not so much in revolutionising fashion as in marketing it. Worth introduced the designer label and the idea of using live mannequins to model his creations. These innovations helped to make him the first fashion tycoon.

DESIGN ICON

It was a woman designer, however, who realised that couture clothes could be both elegant and comfortable. Seamstress and milliner, Gabrielle 'Coco' Chanel (1883–1971) opened her first dress shop in Paris in 1910, but was forced to close her second during the First World War. She reopened in 1919 and by 1924 was leading the fashion world. Taking her inspiration from men's clothing, she introduced a relaxed way of dressing for women based on wool jersey or cotton dresses, trousers, cardigans and costume jewellery. Practical, modern and stylish, her designs revolutionised women's fashion and were soon copied the world over.

Next to the skin

'In olden days a glimpse of stocking was looked on as something shocking. Now, heaven knows, anything goes.'

'ANYTHING GOES' BY COLE PORTER, FROM THE SHOW OF THE SAME NAME, 1934

Many items that would now be considered undergarments were once outerwear. In ancient Egypt the basic and often the only item of clothing for men was the linen *schenti* or loincloth. Children and servants often went naked, while dancers were scantily clad in girdles. Higher ranks wore draped robes in public, but, like the Greeks, they did not distinguish between outer and undergarments.

Nevertheless, it was the Greeks who, in about 400 BC, introduced the forerunners of the bra and corset, although neither seems to have been common. A breast band flattened the bust, while the figure was shaped by a *zoné*, a band of linen or kid worn around the waist and lower body.

Roman women wore similar items under their tunics. For Roman men a linen loincloth was the original and sole undergarment. During the winter, emperors wore *femoralia*, knee-length drawers; these were adopted by soldiers during the 2nd century BC and by civilians in the following century.

Then, in the 4th century AD, the loincloth was replaced by the *camisia*. For men it took the form of a loose shirt and for women a longer smock or shift. Later known by the Norman term *chemise*, this was the essential undergarment for both sexes until the 19th century.

By the Middle Ages, influenced by Germanic tribes who overran Europe after the fall of Rome, men's outerwear consisted of a short tunic and breeches called *braies*. But with the Norman fashion for longer tunics, braies became undergarments, growing shorter and shorter.

Hosiery history

As braies shrank, leg coverings grew longer. Roman *fascia* were lengths of cloth wrapped around the leg, but were thought effeminate and suitable only for old people. Nevertheless, it was men who first wore knee-length stockings, or *chausses*, in western Europe in the 9th century. From about 1340, when they were renamed 'hose', meaning 'leg covering', they fitted the leg tightly from foot to crotch. Women, too, began to wear stockings, but until the 1960s only ballet dancers and actors wore tights. In the 16th century men's 'netherstocks', covered the lower legs and became known as stockings.

The slim waist became the foundation of European fashion in the 14th century. At first the new look was achieved by lacing, but the Elizabethan 'body', from which the word 'bodice' derives, was made of rods of bone or wood. 'Body' gave way to 'stays' in the early 1600s and to 'corset' – from the Old French *cors*, 'body' – in the late 1700s. By then it was a separate garment, usually made of whalebone, rather than part of the dress itself.

Respectable Victorian women were weighed down by layers of underwear, but in the 1880s concerns about health and hygiene brought innovations such as the German doctor Gustav Jaeger's woollen underwear, which allowed the body to 'breathe'.

With the 20th century, underwear became lighter and simpler. In 1935 the American underwear manufacturers Coopers patented the Y-Front. At first called 'Brief Style 1001', it was inspired by a magazine photograph of a man wearing a pair of swimming trunks on the French Riviera. On its launch in Chicago, the entire stock of 600 pairs sold out on the first day. Y-Fronts were made in Britain from 1938, but by the late 1940s were facing competition from boxer shorts, worn as US army issue during the war.

Hidden support

For women, the revolution began around 1910 when the S-shaped figure of Edwardian fashions gave way to a more natural line. Even so, early brassieres – the term is first recorded in American *Vogue* in 1907, although why a French name for a baby's vest was adopted is unclear – were rigid affairs that also covered the midriff.

But in 1913 an American debutante named Mary Phelps Jacob (later Mrs Caresse Crosby) rebelled. One evening before a dance she asked her maid to tie two handkerchiefs to a length of ribbon, thereby creating the modern brassiere. Jacob's friends loved the idea, which she patented in 1914, but after failing to market it successfully she sold her patent to Warner Bros for $15,000. With the introduction of synthetic fibres, the reign of restricting steel, whalebone and canvas finally came to an end.

In the late 1700s women adopted chemise-like gowns for sea bathing, then newly fashionable. Most men swam naked until the mid 1800s, but in 1870 the all-in-one suit appeared. A knee-length skirt made the one-piece costume acceptable for women by 1914, but by the 1920s it reached the hips only. In 1946 the French couturier Jacques Heim designed a two-piece outfit whose impact was compared with the atom-bomb tests on Bikini Atoll.

Everyday hats
Headwear for all occasions

Animal skins, leaves and straw made the earliest headgear, while animal skulls covered in fur became the first helmets. These hats afforded protection from the weather or from aggressors, but they were also status symbols. In both ancient Greece and Rome, only freedmen were entitled to wear them.

Being easy to make, the cap was worn in most early civilisations. The ancient Greek model was the close-fitting felt *pilos*, an early type of beret. It was the women of Minoan Crete who, around 2100 BC, sported what may have been the first 'smart' hats: pointed hats, berets, turbans and three-cornered constructions with ribbons or plumes. But from about 1200 BC feminine headwear consisted of scarves, hoods, veils or headdresses – hats for women did not reappear until the 16th century.

During the 16th century a soft black cap was regulation wear for tradesmen and artisans. Gentlemen adopted the flat peaked cap for country pursuits in the late 19th century, but made sure never to wear it 'in town'.

From Roman times until the Middle Ages most men wore caps or hoods. Ways of draping hoods grew increasingly intricate, however, and by the early 1400s the 'fashioned hood' was in effect a hat, kept in shape with a padded ring and wicker hoops. Thereafter, men of fashion donned hats.

Both sexes wore bonnets in the 1500s, but these soft, velvet affairs bore no relation in style to the rigid bonnets women adopted in the Victorian era. There was also an Elizabethan bowler with a hatband made of crepe, silk and pearls, or silk and spangles.

BOWLED OVER

The traditional men's bowler was designed by the London hatters Lock's of St James's in 1850, at the request of William Coke, a Norfolk landowner who wanted a hard, close-fitting hat for his gamekeepers. It became known as a bowler after Thomas and William Bowler, the firm of feltmakers who manufactured the hat. The grey, flat-brimmed version named after the Earl of Derby caught on in the USA, where men's clothes were more informal.

The other great hat of the 1800s was the top hat. Until about 1830 it was known as a high-crowned beaver hat, after the material from which it was made. The silk top hat was invented by a London hatter, John Hetherington. In 1797 he provoked a riot and was charged with 'a breach of the peace for … wearing upon his head a tall structure having a shining lustre and calculated to frighten timid people'. But by 1850 the topper was being worn by all classes.

From the 1870s smart Englishmen sported Panamas, made since the 1600s from strands of leaves from a palm-like plant in Ecuador and Colombia. Men and women adopted the boater in the 1880s, while in the 1890s the stiff felt homburg was popularised by the Prince of Wales, one of whose favourite spas was the German town of Homburg, where the hat was made. A new style made its debut in George du Maurier's 1895 play *Trilby*, in which the actor Beerbohm Tree played the character of Svengali in a soft-framed black hat.

Boots and shoes
Fashion for the feet

Hunters were probably the first people to think of protecting their feet by wrapping them in animal skins – the earliest surviving examples date from the 2nd millennium BC. A Spanish cave painting of 13,000 BC depicted both a man and woman wearing boots of fur and animal skin. Gradually, however, boots became the essential footwear for men, while leisured women wore decorative but impractical shoes. Only in the 1830s did ankle boots – considered suitably modest – become fashionable for women.

In the 1820s the hero of Waterloo, the 1st Duke of Wellington, gave his name to a tall, slim-cut leather boot. It reigned supreme until the 1860s, when it lost out to the elastic-sided boot, invented in 1837 by Queen Victoria's bootmaker, J Sparkes Hall, and reinvented in the 1950s as the Chelsea boot.

During the pre-Roman era, northern Europeans made shoes from a single piece of animal skin. They pierced the edges with holes, then threaded through a leather thong. This was pulled tight and tied on top of the foot. The Native American buckskin version, the *maxkeseni*, came to England in the 17th century as the moccasin and was briefly in vogue until it fell victim to an import tax.

The brogue, too, started out as a rough shoe of undressed leather tied on with thongs in 16th-century Ireland and Scotland. Holes allowed bog water to drain out while the wearer was walking. The shoe proved so practical that British gamekeepers and their masters adopted it, and by the late 1800s the shoe was heavier and square-toed. It became more refined until, in the 1930s, the Prince of Wales shocked conservative society by wearing suede brogues with a lounge suit.

HEIGHTS OF FASHION

In the late 11th century shoemakers began to stitch together the sole and upper: previously shoes had been made in one piece. Silk and velvet began to be used, while buttons and buckles became alternatives to laces when fastening footwear.

From the early Middle Ages wooden clogs or pattens protected feet from Europe's filthy streets. In the 1500s the Venetian *chopine* had a wedged sole that raised its wearer as much as 50cm (20in). Later in the 16th century a more practical solution was made by building up layers of leather to form a wedge at the end of the sole, but by 1600 shoes had recognisable heels. As a result it became more economical to make 'straights', which were worn on either foot, and left and right shoes disappeared until the late 18th century.

Both men and women wore heels. The French king Louis XIV added an extra 12.5cm (5in) to his small stature, while ladies at the court of Louis XVI could not tackle stairs unaided. Heels did not reach such extremes again until the stiletto appeared in Italy in 1953.

GAINING AN ADVANTAGE

✳ In ancient Greece, courtesans wore sandals with nails studded into the sole so that their footprints would leave the message 'Follow me'.

✳ Roman charioteers wore the first peaked caps – made of bronze – to shield their eyes from the sun's glare as they raced round the track.

✳ The expression 'hat trick' dates from a cricketing practice of the late 1850s, when bowlers who got three batsmen out in successive balls were rewarded with a new white hat.

Jewels and jewellery
The art of adornment

Long before they wore clothes, people adorned themselves with necklaces bracelets and pendants fashioned from seashells, bones, pebbles, mammoth tusks or wood. These objects were valued not only for their beauty, but also as status symbols and amulets.

In western Europe the idea of ornaments for the body began around 38,000 BC. An industry in bone and ivory that was flourishing 4,000 years later in France produced pierced or grooved animal teeth and ivory rings. By around 5000 BC the jewellery-makers of northern Iraq were able to drill the tiny holes necessary to make necklaces out of obsidian, cowrie shells and stone.

Gold was probably discovered in Mesopotamia before 3000 BC, and the jewellery of Sumeria is among the most extraordinary ever made. When the Sumerian queen Pu-abi died in about 2500 BC, she was buried with a cloak of beads made from gold, silver, lapis lazuli, carnelian, chalcedony and agate. On her head gold flowers rose above three diadems. The Sumerians mastered many of the techniques still used by jewellers – metalworking, filigree, stone-cutting and enamelling. But it was the Egyptians who perfected the art of making glass beads, in the 3rd millennium BC. They were first manufactured for a commercial market in Egypt around 1400 BC.

PRECIOUS GEMS

Turquoise and lapis lazuli were among those gems prized and worn by the ancients, while emeralds were found in Upper Egypt around 1650 BC. Emeralds were first brought back to Europe in the late 1500s by explorers from South America, where they were worn by the Incas.

Diamond deposits, discovered in the streambeds of India, were also known to the ancients. They were valued for their rarity rather than for their sparkling beauty since up until the late Middle Ages no one knew how to cut them – diamonds are 85 times harder than sapphires or rubies. Their full glory was revealed in the 17th century, when a Venetian lapidary named Vincenzo Peruzzi developed the 'brilliant' cut.

Because gemstones were so valuable, the Romans built up a large industry in fakes. Their craftsmen were particularly skilled in making artificial emeralds from coloured glass that had been heated with copper or iron-containing chemicals, but also fashioned artificial rubies, sapphires and pearls. In the 1600s a new method using a brilliant glass was developed in Paris to make gem-like stones. Known as paste jewellery, it came to be prized in its own right, and was the precursor of the costume jewellery of the 20th century.

COOL ACCESSORIES

Feathers attached to a handle were used as fans around 3000 bc by the Chinese. They soon created flat hand fans by stretching silk, bamboo or palm leaves over wooden frames. The Japanese discovered in the 6th century ad how to make folding fans. They used painted or embroidered paper, silk or lace for the mount, and ivory, carved wood or mother-of-pearl for the blades. In the 16th century such folding fans became status symbols for women in Europe, where feather fans had been used from the Middle Ages. Feather fans remained popular: ostrich fans were fashionable in the 19th century.

Walking sticks and umbrellas
Style in the hand

Prehistoric chiefs carried with them sticks decorated with emblems, or staffs with carved antlers. In ancient Egypt sticks of various forms were carried by everyone from shepherds to pharaohs. Swordsticks, with a blade hidden in the hollow handle, were among the 132 sticks found in Tutankhamun's tomb dating to 1325 BC.

On a monument built in 2400 BC the Assyrian king Sargon of Akkad is shown being protected from the sun by a servant holding a parasol. The Roman *umbraculum*, meaning 'shady place' or 'bower', was a sunshade made of cloth stretched over a wooden frame. The idea of a device for protection against the rain first appeared in China, where silk umbrellas were used by the nobility from 1000 BC. Waterproof versions appeared between the 4th and 6th centuries AD, constructed of paper made from mulberry bark and then oiled.

The modern umbrella was first seen in Italy in the late 16th century, and spread to France and England in the 1630s. The English adapted the Italian name *ombrella*, 'little shadow', but the French chose the more accurate *parapluie*, 'against the rain'. Initially umbrellas were made of leather or heavy oiled canvas, supported by whalebone ribs, but in 1829 a Parisian factory began to manufacture fine umbrella silk. Henry Holland of Birmingham patented the first successful metal ribs made from steel tubes in 1840.

Purses and bags
For personal possessions

Until the Middle Ages personal effects were wrapped in a piece of material, and hidden inside the clothes. By the early 13th century this had developed into a pouch, bag or purse made in leather, fur or cloth.

Bags were larger than pouches or purses, but all three were worn attached to the belt or girdle by a cord or thong that could be cut from behind – hence the expression 'cutpurse' for a thief. In fact, purses rarely contained money since few people had any to carry. Men used them for their documents and women for sewing implements.

In the 16th century men's hose became so voluminous that vertical pockets could be sewn into the seams. The suit, introduced in the late 1600s, had horizontal pockets in the waistcoat and coat, making men's bags things of the past until the 1830s, when the briefcase appeared. The purse, small enough to be carried in the pocket, was used only for coins.

In the mid 17th century women wore cloth bags fastened to their undergarments. Access to these was by way of slits in the skirt. With the slimline Empire styles of the late 18th century, however, French women carried small bags called reticules. The modern handbag, in the form of a framed leather bag, was born in the 1850s. But it was not until tighter skirts came in during the 20th century, that the handbag came into its own.

PROBABLY THE FIRST MAN IN LONDON TO CARRY AN UMBRELLA REGULARLY WAS THE ENGLISH TRAVEL WRITER AND PHILANTHROPIST JONAS HANWAY. FROM 1750 HE DEFIED HIS CRITICS – WHO THOUGHT IT WAS THE GODLY PURPOSE OF RAIN TO MAKE PEOPLE WET – FOR SOME 30 YEARS BEFORE HIS HABIT CAUGHT ON.

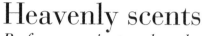

Heavenly scents
Perfumes ancient and modern

The ancients used perfume mostly in the form of incense, a practice reflected in the origins of the word, from the Latin *per*, 'through', and *fumus*, 'smoke'. On public occasions, Egyptians placed on their heads cones of scented fat that melted in the heat, perfuming their hair, faces and bodies.

Although the Egyptians began to distil perfume from lilies around 1350 BC, the art of distilling rosewater and other floral scents was perfected in the Islamic world in the 9th century AD. By the late 1100s perfumes were a central part of European culture, but were not widely used in England until the 1500s. Sweet lavender water, distilled in Surrey from the early 17th century, was Britain's first commercially produced scent.

In the early 1700s, two Italian silk dealers based in Cologne began to sell a family recipe for aqua admiralis, a refreshing blend of oils of neroli, rosemary and bergamot. Renamed eau de cologne in 1709, its fame spread during the Seven Years' War of 1756-63 thanks to its popularity among the soldiers who were stationed in the city.

One of the first perfumes to contain synthetic odours was created in 1923 by the Frenchman Ernest Beaux. He submitted samples to Coco Chanel, who chose the fifth, naming it Chanel No.5, allegedly after her lucky number.

Cosmetics and make-up
The changing face of beauty

When prehistoric peoples painted their bodies their intent was not mere ornamentation: a decorated body was believed to give protection from the forces of evil. From around 4000 BC, Egyptians wore cosmetics as a defence against eye diseases and the blazing sun. The black paste kohl was applied around the eyes and to the lashes and brows.

In the 14th century BC the Egyptian queen Nefertiti painted her nails red with henna, but it was the ancient Chinese who created nail polish from gum arabic, egg white, beeswax and gelatine. A lipstick from a Babylonian tomb of about 4000 BC is the oldest cosmetic found in the Middle East.

Roman women ensured a lily-white complexion by using either chalk or a poisonous cream of white lead, developed in the Indus Valley in the 3rd millennium BC.

Until the 18th century cosmetics were the preserve of royalty, the wealthy and prostitutes. Only in the 1880s did firms such as Eugène Rimmel and Boots make face preparations more widely available. Painted nails became all the rage after the introduction of the first liquid nail polish in 1913.

THE TERM 'MAKE-UP' WAS USED ONLY FOR THEATRICAL PRODUCTS UNTIL IT WAS MADE RESPECTABLE FROM 1920 BY MAX FACTOR, A POLISH IMMIGRANT WIGMAKER. AFTER EMIGRATING TO THE USA MAX FACTOR DEVELOPED FILM MAKE-UP IN 1914. IN 1918 HIS 'SOCIETY' RANGE BECAME THE FIRST MAKE-UP FOR THE GENERAL PUBLIC.

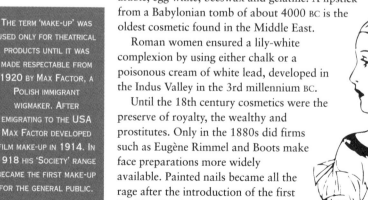

CHAPTER 5

TOOLS AND TECHNOLOGY

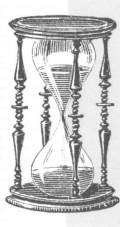

When Louis XV had the first lift built, wanting to find an easier way to reach his mistress's second-floor bedchamber in the Palace of Versailles, he unwittingly started a technological revolution that would shape the skyline of today's cities. More than a century later, in 1854, Elisha Otis developed the first safe and speedy elevator, which in turn made it possible to build much higher than simple staircases had previously allowed. This, combined with the use of concrete and steel, led to the development of modern skyscrapers.

Man has been using such ingenuity for more than 2 million years, ever since someone first picked up a stone and used it to chip at another one. The first tools helped man feed and defend himself, but technology really took off when nomadic peoples settled down to farm. They exploited the natural materials they found – such as wood, stone and clay – to make implements. They harnessed the power of water and wind to drive early labour-saving machines that could grind grains or raise water from wells. And they unlocked the Earth's secrets by mining for metals and minerals. So when metal was first smelted, in around 5000 BC, or electricity first generated, in the 1830s, or fibre optics invented, in the 1950s, ingenious man was simply carrying on the tradition of transforming his world with the wonders of technology.

The wheel
Moving forward

It is difficult to think of a world without the wheel, but 5,000 years ago the pharaohs of Egypt had no knowledge of it when they built the great pyramids. And about 1,500 years later the ancient Britons who constructed Stonehenge used wheel-less technology.

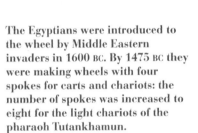

Wheels can easily be imagined to have originated from rollers – tree trunks used to move heavy loads. However, a slice cut from a tree and used as a wheel would have been weak and quick to split. The first wheels clearly recognisable as such appeared about 3500 BC in Sumeria, made from three planks of wood clamped together with wooden struts. Dating from about this time, a sketch on a clay tablet survives from Uruk in Mesopotamia and shows a sledge adapted for easier use on varied terrains by the addition of four wheels. Simple potters' turntables were also employed in Mesopotamia during this period.

In central Asia by 2500 BC, and in Crete by 2000 BC, copper nails were hammered in around the circumference of a wheel to improve its wear. Ancient spoked wheels that were carved from wood usually had iron rims to increase their strength and durability, a basic design used well into the 19th century. Generally, elm was cut for the centre or nave; ash for the rim, and 'Johnny Oak made the spoke'.

The Egyptians were introduced to the wheel by Middle Eastern invaders in 1600 BC. By 1475 BC they were making wheels with four spokes for carts and chariots: the number of spokes was increased to eight for the light chariots of the pharaoh Tutankhamun.

Spoked wheels proved too weak to bear heavy steam-driven vehicles in the 19th century. During the 1820s a steam-bus operator, Walter Hancock, removed the nave and shaped the ends of the spokes into wedges that fitted together at the centre. This stronger 'artillery wheel', named from its adoption for gun carriages, was used in the first cars of the 1880s.

Two wheels with teeth or cogs around their edge, were the gears first employed in around 200 BC in North Africa, for raising pots of water from wells. In the 4th century BC the Greek philosopher Aristotle mentioned what seem to be gears, and 150 years later Ctesibius of Alexandria used a toothed rack and wheel (a rack and pinion gear) to turn the cogs in a water clock.

From lever to pulley
Prime movers

Using simple levers, Stone Age people were able to move heavy loads long before the wheel was ever invented. In the 3rd century BC the Greek mathematician Archimedes explained how levers worked when he showed that a simple lever multiplied force in proportion to its length on either side of its fulcrum, or its point of support. The potential power of the lever was, Archimedes realised, almost infinite: 'Give me but a place to stand and I will move the Earth,' he declared.

The Greeks also used the winch, a combination of a roller and a lever, to raise heavy weights during the construction of such great buildings as the Parthenon. A rope wound around a roller exerted a greater pulling force, they discovered, when the roller was fitted with long arms to turn it. In ancient building work, winches were often driven by men walking inside a vertical wheel which turned the device.

Cranks were used by the Chinese as winding handles on winnowing machines some 2,000 years ago, and in the Middle Ages they were employed for turning grindstones. In the early 17th century, cranks with connecting rods, or crankshafts, began appearing in hand-operated mills and pumps.

The first mention of a true pulley – essentially a wheel with a grooved rim through which a rope or chain is pulled to lift heavy objects – occurs in the *Mechanica*, which was written in Greece in the 4th century BC, while compound pulleys, or block-and-tackle, were described before AD 100 by the Greek scholar Hero of Alexandria. Like the lever, the compound pulley multiplies force while proportionally reducing the distance that objects move. The Romans used compound pulleys in cranes, and 15th-century sailing ships had hundreds of them for hoisting and trimming sails.

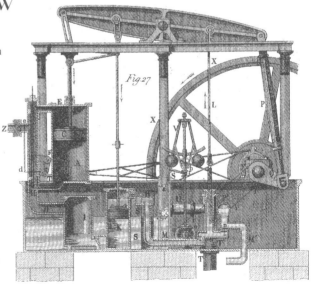

> ## ENDURING TECHNOLOGY
>
> ✣ The power of scissors, pliers and nutcrackers depends on the lever, as does that of the simple winch and the oar.
>
> ✣ A crane combines the principles of both the pulley and the lever.
>
> ✣ The axe and chisel are wedges that are used to split wood. Fitting an axe head to a handle increases the force applied to the wedge.
>
> ✣ Nails are a type of wedge that are used to hold pieces of wood together.

Wedge and screw
Raising agents

Stone Age people used wedges more than 300,000 years ago to raise boulders. Much later, Egyptian quarrymen hammered home wedges to split huge blocks of stone. Axes, first shaped in flint more that 250,000 years ago, use the principle of the wedge.

A screw thread works as an extended wedge. Continuous spiral screws, described by Archimedes in the 3rd century BC, were used for pumping. But the Archimedean screw was old by the time its namesake noted it, having been invented in ancient Egypt. The Greeks and Romans employed the screw to press grapes and olives.

Fig 27

Making pottery
Fired earth

People living in caves in Kyushu, Japan, around 10,000 BC discovered that when clay was heated in fire it formed a hard waterproof material. So began the recorded history of practical pottery, although small clay figures in animal shapes found in eastern Europe may have been made for religious purposes as long ago as 25,000 BC.

As hunter-gatherers across the world settled down into farming communities, they needed vessels to store water, oil and food. By 7500 BC pottery, which was then well established in China, was becoming widespread in both Africa and the Middle East.

The first pots were made by shaping the clay by hand. Pottery could be left in the sun to harden, but it was stronger if placed in a fire. By 5000 BC the kiln, a clay oven heated by fire, had been invented in Asia. Its higher temperature produced a molten paste that ran into any cracks, with the result that pots became more waterproof.

Because most clays contain iron, which turns rust coloured when heated in air, pots are naturally red. The Egyptians added charcoal and oil to clay to make black pottery while the Mesopotamians covered their pots with copper ore before firing to make them blue. Glazes were created around 1600 BC, when potters found that ores of tin produced a white finish. Using a lead glaze that turned to glass during firing made pots even more waterproof.

The first potter's wheel, used around 3500 BC in Mesopotamia, was a simple turntable. In 1500 BC the Greeks attached a disc to the base of the turning shaft that allowed the potter to turn the wheel with his feet.

FINE CHINA

It was the practical genius of the Chinese that gave us porcelain. Using kaolin, a clay that forms a glassy material when it is heated, together with a feldspar known as petuntse, and firing at temperatures of 1,400C (2,552F), potters of the early Tang dynasty (AD 618-906) produced a translucent, hard, white material that had a ring like a bell. Arriving in Europe in the 16th century aboard Portuguese ships, porcelain was first called 'China-ware', or Chiney, from the Persian word *chini* and later became known as china.

A soft-paste porcelain, so called because it could be fired at lower temperatures than true porcelain, was developed in Italy in 1575 by mixing kaolin with glass. But it was not until 1707 that Johann Böttger, alchemist to Augustus II of Saxony and Poland, worked out how to make the real thing. The first true Western porcelain was made at a factory set up by Augustus at Meissen, near Dresden, in 1710.

Bone china appeared in the 1750s, when potteries at Bow and Chelsea added bone ash to soft paste. Easier to make and stronger than true porcelain, it became the standard English porcelain. At the same time Josiah Wedgwood began producing a lead-glazed white earthenware. From then on his Staffordshire potteries turned out high-quality goods at affordable prices.

CHINA
GLASS
AND
QUEENS WARE

The story of glass
A brilliant discovery

A happy accident probably gave us glass. Around 3000 BC Middle Eastern nomads camped on the sandy shores of a lake discovered tiny droplets of glass in the places where they had lit their fires. What they did not realise was that the glass had been formed by chemical and physical changes created when sand (silica), containing significant amounts of soda (sodium carbonate) deposited from the waters of the lake, was heated.

Glass beads were used as common currency in the ancient world. In about 1500 BC the Egyptians made hollow glass by forming a core from a bag of sand or a lump of clay, attaching it to a metal rod then covering it in molten glass. When the glass had cooled the bag or clay was removed.

Glassworking became simpler after about 300 BC when, in Alexandria, molten glass was poured into a mould, a second mould pressed into it and the glass shaped between the two. Around 30 BC a glass-maker in Syria, then under Roman rule, discovered that a blob of glass on the end of a tube could be blown into a vessel of almost any shape, then manipulated to create handles or spouts. The Romans then developed the technique to mass-produce a wide range of glass objects.

Flint, bone and wood
The earliest tools

By hurling weapons made of sharp flint or bone at ibex, mammoths and bison, the people of the Stone Age brought down their prey some 500,000 years ago. With tools of the same natural materials, flesh could be cut from carcasses, hides stripped off for leather, and antlers and tusks removed for fashioning into missiles and cutting implements. Animal sinew was also used to lash pieces of wood together for spears.

Some 400,000 years ago Europeans carved wood into spears, but it was the development of chisels in Europe in around 8000 BC that made true joinery possible. Carved wooden handles for the axe and adze (a tool like an axe but with a curved blade at a right angle to the shaft and swung between the legs) were made in the Middle East from 5500 BC. Adzes were the basic tools for shaping wood until the 1700s.

By 2500 BC the ancient Egyptians were producing richly carved and decorated furniture inlaid with ebony and ivory. Using lathes, which had already existed for nearly 1,000 years, the ancient Greeks created turned furniture. The lathe was used to rotate the wood steadily against a cutting tool, such as a chisel, held by a craftsman. In the design devised by the Greeks for making bowls and vases, a cord attached to the top of a springy vertical pole was wound around the wood, which was spun by a craftsman operating a treadle to pull the other end of the cord. When the treadle was released, the pole spun in the opposite direction.

From copper to bronze
Precious and practical

More than 9,000 years ago the people of south-eastern Turkey discovered that copper could be cut from the shiny rocks of the region with stone tools. However, the word copper is a corruption of the name Cyprus, which supplied much of the metal to the ancient world.

Pieces of copper were soft enough to be beaten and shaped into pins and other jewellery, while malachite, a bright green copper ore, was a perfect material for making beads. Copper-working has continued uninterrupted since these early times, but the metal's value appreciated considerably in the 17th century when it was found to be an effective conductor of heat, and yet more so when, in the 19th century, experiments with electricity proved it to be the most efficient conductor of electric currents after silver.

So revered was their craft that Egyptian metalworkers were probably treated like magicians. The processes that they used, such as forcing the forge fire with blowpipes, were enshrined in the reliefs, hieroglyphs and wall paintings created to decorate ancient tombs.

But it was lead that spurred the major breakthrough in metal production. In about 5000 BC, at both Çatal Hüyük in Anatolia and Yarim Tepe in northern Mesopotamia, rocks rich in lead ores were smelted (heated so that the ores convert to pure metal) and the resulting lead made into necklaces and bracelets.

Tin was extracted in Egypt from about 3500 BC. Smelted with copper over wood fires, it was used to create bronze, an alloy ideal for hard-edged axe and spearheads. Bronze objects, from weapons to jewellery and ornaments, are highly resistant to decay and are still found in ancient sites around the world. They were included in collections of goods buried with the bodies of the rich and powerful; the tomb of Tutankhamun, which was completed in about 1350 BC, contained bronze knives, pins, razors, swords and a bronze trumpet.

For many centuries bronze objects remained more precious than practical, probably because stone was more available and easier to work, but around 1700 BC the people of the Aegean discovered that bronze did not always have to be heated before it could be worked. By 'cold hammering' alloys containing about 10 per cent tin, they crafted a wide range of tools and weapons.

Iron and steel
Industrial production

Pure iron came from outer space. The ancient Egyptians and Aztecs first encountered it in fallen meteorites. In Egypt they called it 'the metal from the sky', and in Mexico 'the gift of heaven'. Since iron usually occurs on Earth in impure ores, meteoritic iron was considered more valuable than gold.

The Iron Age began when, around 1500 BC, the Hittite people of western Asia occupied Anatolia and discovered a way of smelting iron from local ores. They built clay-lined furnaces, and put iron ore and wood in them. On firing, the wood turned to charcoal, and the carbon in it combined with the iron ore to produce metal mixed with 'slag', a glassy residue. Repeated hammering beat most of the slag out of the mixture, leaving relatively pure iron which was then 'wrought', or beaten when hot, to produce blades. Something close to steel – iron containing less than 1 per cent of carbon – was created when charcoal was hammered into the surface. The Hittites called it 'good iron'.

Cast iron was made by the Chinese around 600 BC by melting iron containing phosphorus, which liquefied at a lower temperature than other iron ores. The molten metal was then poured or cast into moulds. This technique transformed Chinese life by making possible the production of such implements as ploughshares and cooking pots.

In the West, the first cast iron was a 14th-century material. A blast of air from a furnace, called a 'blast furnace', heated it to a temperature of 1,200C (2,190F). When fired with charcoal the iron absorbed significant amounts of carbon, which lowered its melting point. Cast into cannons, medieval iron changed the face of warfare. In Britain iron from blast furnaces was known as pig iron because, cast from large troughs into smaller moulds, it resembled a sow feeding her piglets. By the 17th century cast iron was being used for water pipes and a century later, during the Industrial Revolution, it was fashioned into railway tracks.

METAL MANUFACTURE

Until the 18th century, vast areas of woodland were felled to make charcoal for iron production. In 1709 an English iron manufacturer, Abraham Darby of Coalbrookdale, turned coal into the more efficient coke by heating it in an airless chamber. He then used the coke to smelt iron. The results were increased production and less pollution.

Pig iron was too brittle for most heavy industry, but strong wrought iron was needed for the steam-driven machinery, railways and bridges of the new industrial landscape. In 1784 the English iron manufacturer Henry Cort invented the 'puddling process', a faster method of turning pig iron into wrought iron in a specially designed furnace. Although the first iron bridge had already been built over the River Severn at Coalbrookdale in 1777–81, it was the puddling process that made possible widespread iron construction.

SPOTLESS REPUTATION

Stainless steel was invented in 1913 by Harry Brearley, a Sheffield metallurgist and cutlery maker, who discovered that any steel containing 15 per cent chromium does not rust. Brearley made Sheffield synonymous with stainless steel, although not without such opposition as the comment that 'rustlessness is not so great a virtue in cutlery, which ... must be cleaned after each using.'

Mass-production of steel, stronger and more malleable than iron, began after the English inventor Henry Bessemer devised a way of converting pig iron to steel in his 1856 Bessemer converter. By oxidising the carbon in iron and burning it away, mild steel (iron containing about 0.25 per cent carbon) could be manufactured.

Specialist steels were devised in the 20th century. The car manufacturer Henry Ford pioneered the use of high-tensile sheet metal – a low-carbon steel combining strength with good welding performance – for car bodies in 1908. From the 1920s this steel was used in refrigerators and washing machines.

Mining for coal
Digging deep

Between 200 and 300 million years ago, plants that had carpeted the Earth died, decayed and were compressed layer upon layer to form coal. This 'fossil fuel' was first found by Bronze Age people across Europe in seams exposed to view by movements of the Earth's crust. They gathered it up by hand or chipped it out with sharp flints.

From the 5th millennium BC coal was being excavated in western Europe with flint axes. Shafts were hacked into chalk and limestone, tunnels or galleries were extended with picks made from flint or deer antlers, and debris was dug out and removed with the shoulder blades of cattle. About 2,000 years ago, coal was also burned in China and India. In the same era the Romans extracted 'the stone that burns' in Britain to heat their villas and to smelt copper, zinc and tin. They even transported it in large quantities to Hadrian's Wall, where, around AD 300, a Roman guardhouse at Housesteads Fort was converted into a coal shed.

In the 12th century 'bell pits', named for their shape, were dug around the bases of shafts sunk down to 12m (40ft) deep, but once their roofs had reached the point of collapse pits would be abandoned. When the increasing prosperity of Elizabethan England led to a rising domestic demand for coal, mining methods advanced in response. Bell pits were dug, then extended into a series of 'rooms', with pillars of coal acting as roof supports. Horses walking in circles to drive a windlass or gin hauled the coal to the surface.

Longwall mining, so called because coal is removed from a long coalface and the adjoining area protected by wooden props, was introduced in the Shropshire coalfield in 1770. This technique avoided the wastefulness of using pillars of coal to hold up the roof.

Gas and oil
Drilling for fuel

The remains of single-celled organisms such as plankton, which thrived hundreds of millions of years ago before being buried beneath sea-floor sediments, form our resources of natural gas and oil. While drilling for brine the Chinese, who first attempted to penetrate the Earth's surface around AD 300, hit gas by accident. Taking advantage of their good fortune they burned the gas, boiled the brine and produced salt. Gas seepages also occurred naturally. To the Greeks and Romans they signified divine intervention and sacred flames for religious rites were lit from them.

Methane gas released during coal mining first led to speculation about its potential for lighting. John Clayton, the rector of Crofton, near Wakefield, noted in the 1680s that when

By using a 1,500-year-old Chinese technique and pulverising the rock in a hole 21m (69ft) deep with an iron tool at the end of a wire, Edwin L Drakes struck oil in 1859. The first true gusher was hit in Texas in 1901. Petroleum refined from crude oil was to revolutionise transport in the 20th century.

he heated coal 'the spirit which issued out caught fire at the flame of a candle'. In 1794 the inventor William Murdoch lit his house in Redruth, Cornwall, with coal gas. Naphtha, an oil burned in lamps, was distilled from coal tar in 1822, and by 1859 was being produced on a large scale from rocks known as oil shales.

From the 9th century BC people of the Middle East burned tarry bitumen, which they found seeping from the Earth's surface. But only in 1857, when the German prospector G C Hunäus struck oil and gas near Hanover, was oil encountered by drilling. In 1859 Colonel Edwin L Drake made a much larger find at Titusville, Pennsylvania.

A simple oil refinery was built in Pennsylvania in 1860, but in 1900 an efficient system was developed to separate useful 'fractions' such as petrol. The demand for petrol, whose usefulness had previously been scorned, increased after 1885 when Gottlieb Daimler adapted the internal-combustion engine to run on petrol and Karl Benz built a three-wheeler driven by a petrol engine.

UNDERGROUND RICHES

✣ Coal from Somerset was burned by the Romans in the 3rd century AD to replace wood to fuel the sacred flame at the Temple of Sulis in Bath. They also took it to their villas in Wiltshire and Buckinghamshire.

✣ Pit ponies were brought into mines from 1749 at Tanfield Moor Colliery, Yorkshire. They replaced women and older children who were known as hurriers, hauling piles of coal on wheeled carts to the base of the shaft.

✣ Offshore drilling rigs, such as those found today in the North Sea, were first used during the 1940s in the Gulf of Mexico.

Precious metals
Gold and silver

At the beginning of time, declared the Greek writer Hesiod, the Olympian gods created men of gold who lived like gods in true happiness. Gold's lustre, and its malleability and immunity to decay, first inspired the peoples of the Middle East about 6,000 years ago. The ancient state that extended from Aswan in Egypt to Khartoum in Sudan was called Nubia, meaning 'the land of gold'. The golden statues in the tombs of the pharaohs, including Tutankhamun, attest to gold's value in decoration and ritual.

The first prospectors found gold in tiny grains deposited on riverbeds. They separated them from alluvial sand by using sheepskins to which the gold adhered. This may have inspired the Greek legend of Jason, who underwent a series of arduous adventures to win the golden fleece and so become king.

When the Roman Empire spread to Spain at the dawn of the Christian era, gold and silver mines began to be sunk. In England, coinage was minted from such metals in the 8th century AD by Offa, King of Mercia. Edward I enacted the world's first consumer protection laws: his hallmarking system of 1300 ensured that precious metals mixed with alloys were not passed off as pure.

Generating power

'It was not until about seven o'clock, when it began to grow
dark, that the electric light really made itself known
and showed how bright and steady it is.'

THE *NEW YORK TIMES*, SEPTEMBER 5, 1882, AFTER ITS OFFICES WERE
CONNECTED TO THE PEARL STREET POWER STATION

At the flick of a switch, electricity lit and empowered homes
and factories, and also made possible new means of
communication and entertainment. The seeds of this
revolution were sown in China, where the magic of
magnetism, the force that was destined to unlock the door
to electricity, was probably discovered in the 2nd or 3rd
millennium BC. The Chinese found that the mineral
magnetite, an iron ore also known as lodestone, has
powerful magnetic properties. The Greeks also knew of
it, and the philosopher Thales thought that 'the magnet
has life in it because it moves the iron'. Because he
experimented with lodestone from Magnesia, Thales
described it as Magnesian, hence the word 'magnet'.

Both the Chinese and the Greeks knew that if amber, a fossilised
resin, was rubbed, it attracted small pieces of material, rather like a
magnet. Centuries later, William Gilbert, physician to Elizabeth I,
experimented with both magnetism and this static form of electricity.
He concluded that the invisible 'effluvium' responsible for both of
them was widespread and gave the name 'electrics' from *elektron*,
Greek for 'amber', to all substances that had electrostatic properties.

Electric experiments

The next challenge was to produce electricity in
larger, regular quantities. In 1660 the German
physicist Otto von Guericke managed to
accumulate electricity in a ball of sulphur by
rotating it against his hand, and the 18th-century
American scientist Benjamin Franklin showed that
electricity would flow through metal rods.

But it was the battery that made electricity
available in a continuous flow. In 1800 the Italian
physicist Alessandro Volta placed a circle of
cardboard soaked in brine between two metal
discs. To his delight a current of electricity flowed
from this 'voltaic pile', and the more layers he
added the greater the current became.

Hans Christian Oersted, a Danish physicist, discovered in 1819 that a wire with a current flowing through it would make a compass needle move – because the electricity had created a magnetic field. In 1821, in the first of many ground-breaking and influential experiments, the English physicist Michael Faraday used an electric current to make a wire move around a magnet that was immersed in a bowl of mercury.

By using electricity to create movement, Faraday had hit upon the phenomenon that became the key to the electric motor. Returning to his investigations ten years later, Faraday used magnetism to create a series of 'momentary currents' that were brought about by magnetism. These currents, which he described as 'induced', were destined to be used in the electricity industry from its earliest days.

Dynamic force

Faraday's discoveries also led to the invention of transformers, which could change the voltage of electricity and make it cheaper to transmit over long distances. In 1831 he designed the first dynamo, in which a magnet is used to convert mechanical energy into electrical energy without the need for a battery. The advances in electricity were adapted for practical purposes. From the 1840s electrical systems were used for sending telegraph messages and, in 1858, for powering arc lamps in lighthouses. But it was not until 1881 that electricity first went on public supply, in Godalming, Surrey. A small generating plant driven by a water wheel on the Wey river provided electricity for street lighting and a few private houses.

A dynamo was installed at Holborn Viaduct in London in 1882, by the company whose American founder, the inventor Thomas Alva Edison, was among the many people instrumental in realising the practical uses of electricity. The dynamo created power to light the electric lamps in a post office, a church and a public house. Later that year a version six times more powerful was installed at Pearl Street, New York.

Power supply

Conventional steam engines were used for the first coal-burning power stations between 1882 and 1884, when the Irish-born engineer Charles Parsons invented a turbine which converted steam into rotary motion by forcing it past a series of fan-like blades. The first power station using steam turbines opened in 1888 in Newcastle upon Tyne.

In 1895 the first hydroelectric power station was completed at Niagara Falls in the USA. This incorporated water turbines modelled on the ones invented by the French engineer Benoît Fourneyron nearly 70 years earlier in 1827 – themselves updated versions of the water wheels which the Romans had pioneered for power production.

Dyes and paints
Fixing pigments

Nature endowed the ancients with a rainbow of colours for painting and dyeing. Plants provided many of them: red from madder, blue from indigo or woad, black from oak apples and myrtle, and yellow from saffron. Wood smouldered to charcoal also made an effective black. Local minerals were another rich source of colour, with reds from iron oxides, yellows from clays or ochres, blues and greens from ground lapis lazuli and malachite, and whites from ground seashells.

The oldest surviving dyed fabrics are Egyptian, dating from 2000 BC, but the invention of dyeing is probably much older. The Egyptians certainly understood the use of mordants, metallic salts used to fix dye to cloth. Indigo alone requires no mordant and is probably the oldest dye of all. Its name comes from the Spanish word for India, where the plant grows prolifically.

RICH COLOURS

At the beginning of the Christian era Tyre in the eastern Mediterranean was famous for its purple dye and its smell. Tyrian purple was made by pounding the bodies of the molluscs *Purpura* and *Murex* and then boiling them for three days in a salt solution. Only part of the mollusc was used. The rest was left to rot – hence the foul smell. Tyrian purple was known to the Minoans from at least 1600 BC. Due to its rarity, it was used to dye the robes of emperors and popes.

Such traditional dyes were used until Victorian times. But in 1856 William Perkin, an 18-year-old student at the Royal College of Chemistry, London, made the first synthetic dye, a brilliant mauve, from coal tar. Perkin's 'mauveine' was such a success that the following year he set up a factory to make it at Greenford in Middlesex.

Other man-made dyes appeared but the most significant was a synthetic version of alizarin, the red pigment found in the root of madder, by Heinrich Caro at the Badische Anilin und Soda-Fabrik (BASF) in Germany in 1869. Caro's new process spelt the end of madder dyes and the birth of the modern chemical industry.

Paint requires both a pigment to give it colour and a 'binder' to make it stick to a surface. Since Stone Age cave 'paintings' were generally applied without using such binders they were, strictly speaking, made with dyes or pigments. In the 3rd millennium BC gum arabic, a resin obtained from the acacia tree, was employed in Egypt as a binder. Egg white or beeswax were also widely used for this purpose until the Middle Ages, but both ingredients made paint expensive.

Paint did not become widely used until the 18th century, when linseed oil from flax and zinc oxide, a white pigment, became increasingly available as an alternative to white lead, which was not only toxic but had a yellowish tinge. Mixed together they composed the new paints. A cheaper, safe and much whiter pigment, titanium dioxide, came into use after the First World War.

Pest control
Protecting crops

In the 8th century BC the Greek poet Homer recommended burning sulphur to fumigate plant pests. And in the 1st century AD the Roman writer and natural historian Pliny the Elder suggested using a wormwood compound to kill the caterpillars of cabbage-white butterflies.

   Most early forms of chemical pest control came, as Pliny had suggested, from substances extracted from plants. In France a solution of nicotine, from tobacco leaves, was used to kill aphids in 1763. Pyrethrum, from a type of chrysanthemum, also proved successful.

   More poisonous insecticides were introduced in the 1880s, including Paris Green, an arsenic compound. DDT, or dichlorodiphenyltrichloroethane, was synthesised in 1939 by Paul Muller, a German chemist.

Artificial fertilisers
Boosting crop growth

Until the 19th century, organic materials such as wood ash, bones, dried blood and sea-bird droppings were the main fertilisers. In 1840 the notion of chemical fertilisers was proposed for the first time by Justus von Liebig, a German chemist. Liebig argued that farmers had to replace those chemicals, particularly nitrogen, phosphorus and potassium, which plants took from the soil. Two years later Sir John Bennet Lawes, an English landowner, produced superphosphate, the first artificial fertiliser, by treating coprolites (the fossilised excrement of prehistoric animals) with sulphuric acid to make them soluble in water.

   Synthetic nitrogen fertiliser was invented in 1898 by Adolph Frank and Nikodem Caro, two German scientists working at BASF. But the decisive development came in 1910 when the German chemist Fritz Haber devised the process, which still carries his name, for converting nitrogen in the air to ammonia, which could be used, in turn, to produce fertiliser. In Britain, Imperial Chemical Industries (ICI) began combining nitrogen, phosphate and potash to manufacture synthetic fertilisers in 1926.

POTENT MIXTURES

✣ The ancient Britons painted themselves with woad, a blue dye extracted from the leaves of *Isatis tinctora*, to scare their enemies.

✣ Greasepaint was invented by Ludwig Leichner, a German opera singer, and independently by Carl Baudin of the Leipzig City Theatre in the 1860s.

✣ In 1885 the weed-killing properties of Bordeaux mixture, a combination of copper sulphate and lime, were discovered. It was used to treat a mildew ravaging the vineyards of Bordeaux and became the first chemical weedkiller.

The story of rubber
Waterproof and flexible

In the 6th century AD the Aztecs of Mexico took a substance they called *olli*, which oozed from the trunks of trees, and made it into bouncy balls. They also

invented the game *tlachtli*, knocking the balls around a long court. In 1530 the Italian geographer Pietro Martire d'Anghiera described rubber in his book *De Orbe Novo* (*From the New World*). 'From one of these trees a milky juice exudes,' he wrote. 'Left standing it thickens to a kind of pitchy resin.' By 1615 Spanish troops in South America were using this 'resin' to waterproof their cloaks. After the English chemist Joseph Priestley noted in 1770 that it was useful for rubbing out pencil marks, it became known as rubber.

Rubber was used in the early 19th century to make a variety of waterproof clothes, but it was prone to become sticky in warm weather and rigid in the cold. The Scottish chemist Charles Macintosh solved this problem in 1823 by placing a layer of rubber between two layers of cloth, thereby inventing the raincoat that still bears his name.

Charles Goodyear, an American inventor, started on a series of complex experiments to harden rubber in 1834. After a succession of failed attempts he accidentally reached his long-sought goal in the early 1840s when he spilt some rubber and sulphur on a hot stove. The substance that resulted was both waterproof and pliant.

Synthetic or Buna rubber, which is made from butadiene, a chemical extracted from petroleum, was created after experiments had been carried out by German chemists in 1927. In 1931 a team led by Wallace H Carothers at the American chemical company Du Pont came up with neoprene, a stronger and flame-resistant synthetic rubber.

The plastics revolution
Shaping objects

The word 'plastic' stems from the Greek word *plastikos* meaning 'fit for moulding'. Materials with naturally plastic qualities, including horn, tortoiseshell and shellac, have been used for thousands of years to make ornaments and the handles of tools. But the modern plastic age began in 1862 when the English chemist Alexander Parkes made imitation ivory combs and hairslides. Parkes first added nitric acid to cellulose, a substance abundant in the woody parts of plants. He then used camphor to shape the resulting material, cellulose nitrate, into a pliable form, which he called 'Parkesine'.

In 1869 a similar product was made by the American engineers John and Isaiah Hyatt to replace ivory in billiard balls. They named their plastic 'celluloid', and commercial success followed, despite the material's flammability, with the manufacture of toys and dressing-table sets. The American businessman George Eastman created a new flexible celluloid film for cameras in 1889.

Unlike celluloid, which contained some natural materials, Bakelite was entirely synthetic. Patented in the USA in 1907, it was the result of the Belgian-born chemist Leo Baekeland's search for a shellac substitute. Baekeland's winning formula was a mixture of the chemicals phenol and formaldehyde.

Bakelite was a resinous material that could be moulded by heat and pressure. It was both durable and popular since, once heat-set, it could not be softened or pushed out of shape. Made into telephones and radios from around 1929, its distinctive look – it was made in dark colours, often blended to create imitations of marble and onyx – helped to define the 'modern' style of the interwar years.

The 20th-century plastics industry was not truly launched until 1922, when the German chemist Hermann Staudinger proved that rubber consisted of long chains of a basic molecule repeated in a structure now known as a polymer. He also discovered that styrene, an oily liquid component of crude oil, could be turned into a polymer when heated. During the 1920s polystyrene was developed in Germany as an alternative to rubber and patented by the company I.G. Farben in 1929. Objects were fashioned by injection moulding it into predetermined shapes.

Perspex was developed in 1930 after chemists working in Britain, Canada and Germany discovered that methyl acrylate could be 'polymerised' to create a clear, strong plastic. In 1936 this Perspex was used in the windshields and cockpits of the new Spitfire aircraft.

Fibre technology
Miracle strands

AN ENGLISH CHEMIST NAMED REGINALD OSWALD GIBSON INVENTED POLYETHYLENE, OR POLYTHENE, IN 1933. EASILY MOULDED AND WATER-REPELLENT, IT WAS USED FROM 1939 AS A CABLE INSULATOR AND IN RADAR COMPONENTS. THE FIRST POLYTHENE HOUSEHOLD PRODUCT WAS A WASHING-UP BOWL MANUFACTURED IN 1948.

In 1963 Royal Aircraft engineers at Farnborough in Hampshire discovered that when fibres were heated their molecules changed into strong chains of pure carbon. The result was carbon fibre, exploited in the construction of lightweight performance cars, tennis rackets and golf clubs.

Strands the diameter of a human hair are the basis of fibre optics. In 1955 Dr Narinder Kapany of Imperial College, London, showed that light could travel down a fine fibre made of two types of glass. The different glasses made the light 'bounce' to the end of the strand. Fibre-optic technology was used in the endoscope, a medical device for looking inside a patient's body, developed by the American surgeon Basil Hirschowitz in 1958. In 1966 the scientists Charles Kao and George Hockham, who were working at the Standard Telephone Laboratories in Essex, proved that such fibres could also be used to carry telecommunications signals.

The microscope
Magnifying the miniscule

Without the microscope the amazing world invisible to the human eye, including cells and bacteria, would still be a mystery. At many ancient sites across the world, including some such as Gordion in Turkey that date back to the 9th century BC, pieces of transparent rock crystal with curved surfaces have been found. But whether these crystals were actually recognised and used as lenses, or merely regarded as decorative jewels, is unknown.

If the ancients had understood lenses they would probably have invented spectacles, but these did not appear in Europe until the 13th century AD. It was the Dutch spectacle-makers Hans and Zacharias Janssen who devised the compound microscope, which combined two lenses, in 1590. However, imperfections in the lenses meant only a blurred image was produced.

Some 90 years later Anton van Leeuwenhoek, the Dutch naturalist, used a single, perfectly ground lens to magnify bacteria hundreds of times. Awed by such sights, the naturalist Henry Power proclaimed:

'In the wood-mite or -louse you may behold
An eye of trellis-work in burnisht gold.'

During the 18th century, lenses gave sharper images as chromatic aberration, the presence of coloured fringes produced by the division of white light into its colours, was progressively reduced. But not until 1830 did the English microscopist Joseph Jackson Lister successfully remove the distortion, which resulted from greater bending of light at the edge of a lens than at its centre, known as spherical aberration. About a century later the electron microscope revealed more detail, providing magnification that was some 100,000 times greater than that of a standard microscope.

THE FIRST IMAGES OF STARS SEEN IN EARLY TELESCOPES WERE SURROUNDED WITH COLOURED FRINGES, WHICH RESULTED FROM CHROMATIC ABERRATION. IN 1668 THE ENGLISH ASTRONOMER ISAAC NEWTON SOLVED THIS PROBLEM IN A DESIGN THAT REMAINS ESSENTIALLY UNCHANGED. HIS REFLECTING TELESCOPE USED CURVED MIRRORS TO FOCUS LIGHT FROM THE HEAVENS.

Telescopes and binoculars
Distance viewing

Credit for the telescope's invention is still contested, but a Dutch spectacle-maker named Hans Lippershey is the most likely candidate. He applied for a patent for his 'looker', consisting of a concave and a convex lens in a tube, in 1608. However, his discovery may have been predated by those of the English mathematician Leonard Digges or the Italian physicist Giambattista della Porta, both of whom hinted at similar instruments in the previous century.

In 1609 the Italian scientist Galileo Galilei made his own telescope, turning it towards the heavens. Having identified the moons in their orbit around Jupiter, Galileo deduced that the Earth was not the motionless centre of the Universe.

Binoculars, first seen in Paris in 1823, consisted of a small telescope for each eye. They were improved towards the end of the 19th century when Ernst Abbe, assistant to the German scientist and instrument maker Carl Zeiss, discovered that by putting two triangular glass prisms in each telescope a magnification similar to that of much longer telescopes could be achieved.

The thermometer
Measuring temperatures

In 1592 temperature was probably first measured when Galileo devised an instrument consisting of a glass bulb attached to a slender glass tube, which he immersed in water. Any decrease in the temperature of the bulb lowered the pressure of the air within and sucked the water higher into the tube. But because it was sensitive to changes in atmospheric pressure the instrument was unreliable.

In 1624 the French scientist Jean Leurechon used the word *thermomètre* to mean 'measurer of heat'. The first thermometer to be sealed from the air, and therefore stay unaffected by atmospheric pressure, was made in 1641, by Ferdinand II, the Grand Duke of Tuscany. Ferdinand also used coloured alcohol, but later in the same century Tuscan scientists replaced this with mercury.

In 1724 the German physicist Gabriel Fahrenheit described a temperature scale. He set zero (at the time the lowest temperature considered obtainable) as the freezing point of a mixture of ice, water and sea salt, and 96 degrees as body temperature, which he worked out by placing the thermometer bulb in either the mouth or the armpit. On the Fahrenheit scale, water freezes at 32 degrees and boils at 212.

In 1742 the Swedish astronomer Anders Celsius suggested a temperature scale in which the freezing point of water would be 100 degrees Celsius and the boiling point zero degrees. This illogical scale was inverted by the French scientist Jean Pierre Christin the next year.

More than a century passed before doctors started systematically measuring body temperature to aid diagnosis. In 1868 a German professor of medicine, Karl August Wunderlich, published *The Temperature in Diseases*, based on measurements of 25,000 patients. His research required great dedication since his clinical thermometer was some 30cm (1ft) long and it took about 20 minutes to register each reading.

But help was at hand. In 1852 a thermometer with a constriction in its narrow glass tube to prevent the mercury falling back was patented and in 1867 Clifford Allbutt improved this with a shortened version. Supported by Wunderlich's data, this thermometer became invaluable to physicians.

ON FULL BEAM

As he sat in a park in Washington DC one morning in 1951 the physicist Charles Townes dreamed up the laser's precursor. Planning to amplify microwaves (electromagnetic radiation beyond the visible spectrum) for use in radar, he decided to use high-energy atoms in ammonia molecules to produce 'microwave amplification by stimulated emission of radiation'. An acronym made its name 'maser'. Theodore Maiman, another US physicist, applied the same idea to light and built the first laser, emitting red light from a ruby crystal, in 1960.

Weights and measures
Setting a standard

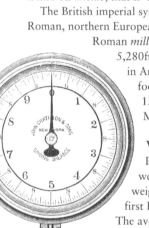

When early peoples needed to measure something, they used their bodies to provide dimensions. The Egyptian cubit, employed from about 3500 BC, was based on the distance from a man's elbow to the tip of his index finger (about 21in/52cm) and subdivided into further units the width of a finger.

By about 1000 BC the Greeks were using a modified cubit in which one of the subdivisions was based on the length of the average man's foot – roughly 12in (30cm). The Romans adopted the foot as a unit of measurement, splitting it into 12 *unciae*, that is 'twelfth parts', or inches.

The British imperial system of weights and measures evolved from a mix of Roman, northern European and improvised units. The mile originated in the Roman *mille passus*, or 'thousand paces'. It was set at its current 5,280ft (1,609m) in the late 16th century. Yards originated in Anglo-Saxon times. The current lengths of the yard, foot and inch were roughly established by the late 1300s and were confirmed in the Weights and Measures Act of 1855.

WEIGHTY MATTERS

Pounds take their name from the Roman *pondo*, a weight of about 12oz (340g), but another Roman weight, the *libra*, is the origin of the abbreviation lb. The first English standard was established in the 13th century. The avoir-dupois pound now in general use, equal to 16oz (453g) and named from Old French *aveir de peis* or 'goods of weight', was adopted in the late 1500s.

The idea of using the dimensions of the Earth as a basis for measurement was first suggested by the French cleric Gabriel Mouton in 1670. Only after the French Revolution of 1789 was a commission established to 'bring to an end the astounding and scandalous diversity in our measures'. The metre, from the Greek *metron*, 'measure', was defined as one ten-millionth of the distance from the North Pole to the Equator.

In 1799 a platinum rod exactly one metre long was established as the standard and deposited in the National Archives. All other lengths were derived from it in multiples of ten, using the

The first standard measurement was set around AD 950 when King Edgar defined the yard as the 'measure of Winchester', a reference to a rod kept at the city that was his capital.

prefixes *kilo* (from the Greek for 'one thousand'), *deci*, *centi* and *milli* (from the Latin words for 'one-tenth', 'one-hundredth' and 'one-thousandth'). The unit of volume, the litre, was set at 1,000 cubic centimetres and the unit of weight, the gram, was defined as the weight of 1 cubic centimetre of water.

The metric system became the French legal standard in 1840 and was soon adopted throughout Europe. In Britain the slow process of changing to metrication began only in 1965.

Calendars and standard time
Counting the days

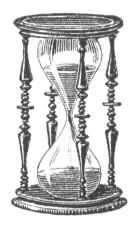

In agricultural societies, the calendar was important because it determined the right time for sowing and harvesting. The farmers of ancient Egypt, who divided the year into three seasons (Nile flood, sowing and harvest), developed a 12-month year based on lunar cycles. But the rhythmic waxing and waning of the Moon does not relate to the solar year, the time the Earth takes to orbit the Sun. By the time Julius Caesar came to power in Rome in AD 46 the seasons were three months out of step with the calendar.

The Julian calendar, devised by the Alexandrian astronomer Sosigenes and named after Caesar, closed the gap by adding an extra day every fourth, or leap, year. This close approximation was still wrong by some 11 minutes a year, and by 1572 this had built up to a 'loss' of ten days. In a revision ordered by Pope Gregory XIII, it was calculated that the value of a solar year differed from the Julian calendar, used throughout western Europe, by just over three days every 400 years. Consequently, in 1582 ten days were 'lost' – October 5 became October 15 – and it was decreed that leap years should occur only 97 times in 400 years; practically, this means that century years are leap years only if they can be divided by 400.

BEFORE AND AFTER

The Gregorian calendar was adopted in France, Italy, Spain, Portugal and Luxembourg, but elsewhere in Europe there was a reluctance to accept a change dictated by the Roman Catholic Church and the new calendar was not introduced into Britain until 1752. The system of consecutively numbering the years of the Christian era, with dates marked BC (before Christ) or AD (*anno Domini*, Latin for 'in the year of the Lord'), was devised in the 6th century.

That the day is divided into two periods of 12 hours each also dates back to the ancient Egyptians. Their 'seasonal hours' varied in length according to the Sun's course, a system still in use in Europe when mechanical clocks became established in the 14th century. Standard time zones were adopted in 1884 by an international conference, which set the zero meridian at the Royal Observatory, Greenwich.

WHATEVER THE WEATHER

Scientific meteorology began in 1854, when the French astronomer Urbain Le Verrier was asked to investigate why a French warship and 38 merchant ships had sunk near Balaklava, Ukraine, during the Crimean War. Le Verrier was able to show that the ships had been engulfed by a 'sudden' storm that had swept across western Europe, thus proving that storms travel around the Earth. This opened the way to weather forecasting of a kind, as did the confirmation by the Dutch meteorologist Christoph Buys Ballot a few years later that windflow follows isobars, lines drawn on charts to connect points of equal atmospheric pressure.

Timepieces
Marking the passing hours

Noticing the way that shadows moved as the Sun advanced across the sky, the Egyptians were inspired to make the first sundials as long ago as 2000 BC. A crossbar was sited so that its shadow moved over a series of marks on the ground as the day progressed.

The Egyptians measured time after sunset by charting the movement of the stars or with water clocks. The earliest known of these clocks dates from the 14th century BC during the reign of Amenhotep III. Standing in the temple of Amun-Re, at Karnak, it consisted of a vessel with a small hole in the bottom through which water dripped, marking time as the liquid's level fell.

Water clocks, or clepsydras (from the Greek word *klepsudra*, meaning 'water stealer'), continued to mark time until the Middle Ages. Many of these clepsydras were ornately engineered, and some sounded bells or moved hands to indicate the hour.

From the 14th century monks in Europe marked time by the hourglass as well as the water clock, rendered ineffective by the freezing temperatures of winter.

Mechanical timekeeping, so written accounts record, began in about AD 1090 when the Chinese astronomer Su Sung devised a water-powered clock controlled by a form of escapement, a notched wheel that was driven by a falling weight on a cord. In Europe the earliest mechanical clocks were made in the late 13th century – one was erected at Dunstable Priory in 1283.

LIKE CLOCKWORK

In the mid 15th century clockwork timepieces driven by the power of a coiled spring appeared. The rate at which the spring turned the driving wheels was kept steady by a fusee, a spiral groove cut on a cone-like form, connected to the mainspring drum by a gut line. Spring power meant that clocks no longer needed external weights and did not have to be hung on walls. This made possible the development of smaller portable clocks and watches.

Galileo was the first to understand how the pendulum worked after seeing lamps swinging in Pisa Cathedral in 1581. But it was not until 1641, the year before his death, that the then blind Galileo asked his son Vincenzio to construct a pendulum clock. Vincenzio never finished his task, and it was left to Salomon Coster and Jan Van Call to produce the first pendulum clocks, to a design created by the Dutch physicist Christiaan Huyghens in 1657.

The anchor escapement appeared around 1670, possibly the invention of the London clock-maker William Clement. The mechanism rocked one 'tick' with each swing of the pendulum, and made accurate timekeeping possible for the first time. In 1675 Huyghens and the English physicist and architect Robert Hooke independently invented the hairspring, a fine spiral balance spring that bought similar accuracy to portable timekeepers such as watches.

THE OLDEST SURVIVING CLOCK IN BRITAIN WAS BUILT FOR SALISBURY CATHEDRAL IN 1386. IT HAS NO HANDS AND STRIKES THE HOURS ON A BELL. INDEED, THE WORD 'CLOCK' COMES FROM *GLOCKE*, GERMAN FOR 'BELL'. SIMILAR CLOCKS, SOME STRIKING THE QUARTERS AS WELL AS THE HOURS, WERE BUILT THROUGHOUT WESTERN EUROPE. BUT THEY WERE NOT ACCURATE, AND A LOSS OR GAIN OF SEVERAL MINUTES A DAY WAS THE BEST THAT COULD BE EXPECTED.

Watches and wristwatches
Portable timekeepers

Until clocks small enough to be carried were invented, sundials were the only portable timepieces. Some of the earliest watches were made at the beginning of the 16th century by Peter Heinlein of Nuremberg in Germany. These watches were housed in drum-shaped cases. Heinlein also made watches that resembled pomanders, small, pierced metal balls filled with sweet-smelling herbs or scent.

Possibly the first watch worn on the arm was 'an armlet of gold, all fairly garnished with rubies and diamonds and having in the closing thereof a small clock'. It was presented by the Earl of Leicester in 1571 to Elizabeth I. A 'watch to be fixed on a bracelet' was recorded in the 1790 accounts of the watchmakers Jacquet Droz and Paul Leschot of Geneva, Switzerland.

However, wristwatches remained extremely rare. Both men and women continued to wear fob or pocket watches. The change began after the German navy issued wristwatches to all its officers in 1880, on the basis that they were easier to consult on a storm-tossed ship. But resistance persisted until the First World War when lightweight and easy-to-read wristwatches proved more practical in the trenches than pocket watches.

PERFECT TIMING

The Rolex company, which was founded in Geneva in 1905 by Hans Wilsdorf, introduced a waterproof watch, the Rolex Oyster, in 1926. Even after it had been immersed in water for three weeks, it kept excellent time.

Electronic watches took the USA by storm when the Bulova Accutron was introduced in 1960. Using a vibrating tuning fork to set the time, it proved to be far more accurate than any mechanical watch. Further innovation followed in 1967 with the development of quartz watches by the Swiss Horological Electronic Centre. Powered by batteries, they contained quartz crystals that vibrated more than 8,000 times a second and made possible the most precise watches to date. Nor did they need to be wound.

Digital watches arrived in 1971. Developed by the American engineers George Theiss and Willy Crabtree, they presented a light-emitting diode, or visual number display, and were called 'Pulsar' watches.

Alarm clocks
Awakened by bells

The Greeks may have been the first to use clepsydras as alarm clocks. Ctesibius of Alexandria invented one clock with ringing bells around 270 BC. From the 6th century AD water clocks served as alarms in Europe's monasteries – the clock-keeper would ring a bell when it was time for the monks to pray. These developed into clocks that would set off the monastery bells automatically. Some of the early mechanical clocks were also used by monasteries to wake monks from their slumbers so that they could pray at matins.

Bridges
Crossing the water

Neolithic settlements were often close to rivers and streams and there was, therefore, a need for safe, dry crossings. The first basic bridges of 10,000 years ago consisted of tree trunks laid across the water, supported by a large stone or wooden pier over wide rivers. Flat stone bridges were constructed for shorter spans.

In the 5th century BC, the Greek historian Herodotus described a bridge he had seen in Babylon with 100 stone piers linked by 1.5m (5ft) wooden beams. For greater spans in stone the Romans were the first to make use of the arch.

Interest in Roman bridges revived during the Renaissance. The Italian scholar and architect Andrea Palladio studied Roman designs before inventing the truss, an open structure of light wooden struts linked by angled crossbeams, in 1570.

The box girder, a variation on the truss, was invented in the 19th century by the Newcastle engineer Robert Stephenson. The girder of his Britannia Bridge of 1850, spanning the Menai Strait in north Wales, owed its strength to being a closed tube, or box, through which trains ran.

The first cast-iron bridge, of 1779, crossed the Severn at Coalbrookdale in Shropshire. Built without any screws, the bridge was held together by dovetail and mortice-and-tenon joints.

The Winch footbridge, the original suspension bridge, was built to span the River Tees in England in 1741. A concrete roadway 21m (70ft) long was introduced in 1801 by the American engineer James Finlay for the Jacob's Creek Bridge in Pennsylvania. The first bridge suspended from steel cables, Brooklyn Bridge in New York, was completed in 1883.

BUILDING ACHIEVEMENTS

✤ By using suspension wires, the world's longest bridges make practicable spans more than 1,410m (4,630ft) long. Their origins lie in Neolithic times when rope or vine bridges were strung across streams.

✤ The Romans first used concrete to build a theatre at Pompeii in 55 BC.

✤ From the 2nd to the 5th centuries AD, Christians in Rome dug tunnels for use as secret underground burial galleries, or catacombs.

✤ The first tunnel beneath the Alps, 73m (240ft) long, was built in 1707. But a major passage under Mont Cenis, measuring 13km (8 miles), was not bored until the 1860s.

Tunnels and tunnelling
Passages underground

The ancient Egyptians constructed tunnels leading to their pharaohs' tombs. The earliest, dug through solid rock in about 1220 BC in the Valley of the Kings, gave access to the tomb of Merneptah (the son of Rameses II). A tunnel 107m (350ft) long led to a shaft opening on to another tunnel of 91m (300ft). At the end of it lay the king's burial chamber. By the 8th century BC Persians and Armenians were digging extensive *qanats*, or irrigation tunnels, to carry water 32km (20 miles) or more from rivers to towns.

Attempts to tunnel underneath rivers proved unsuccessful until relatively modern times. In 1818 a box-shaped iron casing, or shield, was devised by the British engineer Marc Isambard Brunel to support the roof of the first tunnel to be dug under the River Thames. As miners within shovelled through the soft clay the shield was pushed forward and the tunnel gradually lined with bricks.

Skyscrapers
Building upwards

Concrete and steel, which combine strength with lightness, made skyscrapers possible. Equally important was the lift. The first lift had been installed on the outside of Louis XV's Palace of Versailles in 1743 to allow him easy access to his mistress's second-floor apartment. However, safe and speedy lifts were not devised until 1854, by the American engineer Elisha Otis. His lift, supported by a cable, boasted a mechanism that held it in place even when the cable was severed.

After Chicago's great fire of 1871, Otis's invention proved invaluable as the city was rebuilt upwards. The ten-storey Home Insurance Building, completed by William Jenney in 1885, was the first tall structure to use cast iron and steel in a supporting frame. By the mid 1930s skyscrapers had multiplied in American cities. The word itself was far older, having been used since 1794 to describe all sorts of tall items ranging from top hats to ships' sails.

Advances in building
Milestones in construction

The oldest known form of construction is the post and lintel – two vertical supports with a horizontal crossbeam. This probably dates back to prehistoric times, when tree trunks served as posts and branches as lintels. From about 2500 BC the burial chambers within Egyptian pyramids imitated this style in stone, but such monumental structures had limitations because they allowed only short horizontal spans.

Wider openings became possible with the arch, invented by the Assyrians and Babylonians in the 6th century BC. Because a single arch transmitted loads outwards and downwards, it could bear a far greater weight than the post and lintel. It was not until about AD 1050 that the round Roman arch was superseded by the vault, a much steeper arch used to magnificent effect in medieval Gothic cathedrals.

The dome is, in structural terms, a series of arches meeting at a single point. Domes capped tombs from about 2500 BC in the Mesara Plain in Crete, but the world's first great dome, with a diameter of some 44m (144ft), appeared in AD 124 on the Pantheon in Rome. Its size was made possible by a thinly applied layer of concrete, which enabled the surface weight to be minimised.

Water and waste

'It's limpid and clear from all mud
This water I sell for the public good
Its excellent virtues no mortal can tell
So sweet is the water from Union Well.'

DOGGEREL ON THE SIDE OF A 16TH-CENTURY WATERCART,
MONKWEARMOUTH, SUNDERLAND

To preserve and channel precious water, early Bronze Age settlers in Europe encased springs in wooden holding tanks. Where the water table was deep below the ground, they dug wells, lined them with wood, stone or brick, and scooped out the water with hollowed gourds, clay jars or wooden buckets.

The wells of cities of the ancient Near East were deep and sturdy, and one built in the 9th century BC at Nimrud, in modern-day Iraq, still held water when discovered in 1951. In the 1st millennium BC the peoples of this area made cisterns for rainwater, some hewn from stone blocks, others hollowed out from the rocky land.

The Egyptians seem to have thought of damming rivers and wadis, water channels that drain dry outside the rainy season. The oldest surviving dam, a wall of masonry some 80m (270ft) thick and 110m (370ft) long, was built around 2000 BC to contain the spate waters of Wadi Gerrawi, about 40km (25 miles) south of Cairo.

Settlers in the Indus Valley, covering eastern Pakistan and part of western India, were among the first to pipe water into the home. It flowed through clay pipes around 2000 BC. An ambitious water system constructed by the Minoans in Crete at about the same time included overground aqueducts to carry water from the mountains to the palace at Knossos, 11km (7 miles) away. Their terracotta pipes were tapered at one end so they could fit together.

Public waterworks scaled the heights with the magnificent arched aqueducts built by the Romans, but the first of their water channels, the Aqua Appia, constructed in about 300 BC to service Rome, was in fact an underground conduit. This system had first been developed in the Near East, and was used in Persia by the 6th century BC.

Lead pipes were commonly used in Rome, causing some physicians concern about contamination. The Greeks had already experimented with wool, wick and tufa (a porous

PUNCH'S FAN

THAMES

EMBANK

SIR

HE IS GREAT AT DRAI

rock) as water filters, and the Romans added wine to the list of purifiers. Water was also filtered by being percolated through layers of progressively finer sand.

With the fall of the Roman Empire the superbly organised waterworks system collapsed, and plumbing survived in the West only in large public buildings and the great monasteries. Attempts to supply cities were not resumed until 1190, when a network of lead pipes was laid down in Paris. London's first public water conduit was built in 1236; water was taken from the Tyburn stream at what is now Stratford Place in Oxford Street then carried to the City in lead pipes. But such ventures were few until the growth of cities in the mid 18th century led to the creation of private water companies.

TRAITS.—No. 164.

AZALGETTE, C.B.

AS MADE A COMPANION OF THE BATH.

Stone Age sewers

At Skara Brae in the Orkneys, the oldest known British Neolithic settlement, villagers living in simple stone huts around 2800 BC built a drainage system and used rudimentary lavatories. The Greeks and Romans drained waste into cesspits and sewers. The cesspits were rarely emptied, and open sewers gave off such a stench that they were eventually covered.

The same problems affected European city-dwellers up to the 19th century, when the discharge of effluent into streams and rivers, and the overspill of cesspools, led to pollution and the spread of diseases. The first large-scale underground sewers were built in Hamburg, Germany, when the city was rebuilt after a fire in 1843.

London's sewers were redesigned by Sir Joseph Bazalgette, appointed Chief Engineer to the Metropolitan Board of Works in 1845. His sewers, which are still in use, intercepted the city's streams and carried the waste to outfalls north and south of the capital. Pumping stations lifted the sewage up to a higher level to aid the flow.

The deepening of the Fleet Street sewer in 1845 and other efforts to improve London's sewerage system failed to prevent cholera epidemics killing 20,000 city-dwellers. A total overhaul of the system, completed in 1875, led to a dramatic improvement in the health of Londoners.

Papyrus and paper
Writing surfaces

According to legend a Chinese courtier named Ts'ai Lun discovered paper in
AD 105. Noticing scraps of rotting rag and tree bark floating on water, Ts'ai
picked up the pieces on a screen, drained off the water and proceeded to write
on the dried material. In fact, paper may already have been produced for two
centuries, replacing both wood and silk as writing surfaces in China.

Papyrus, made from dried reeds, was used as a writing material in ancient
Egypt from about 3000 BC. Parchment and vellum (made from processed
animal hides) were used in the 2nd century BC at Pergamum, in modern-day
Turkey, and gradually replaced papyrus in Mediterranean countries over the
following centuries. Paper made from linen rags, and hemp and flax cords,
appeared in the Middle East during the 8th century, although parchment
remained the preferred material for religious works and legal documents.
Paper reached Europe from Muslim Spain in the 10th century AD.

Following the fall of the Spanish Umayyad dynasty in the 11th century,
paper continued to be used by the Christian conquerors. As it became more
common across Europe, paper-mills were built, the first in England dating
from about 1494. These mills struggled to obtain enough linen to meet the
demand for paper. Many countries therefore banned the export of this raw
material, and in 16th-century England it was made illegal to bury the dead in
linen shrouds.

With the rise of both literacy and newspapers, the demand for paper
increased further in the 17th century. The continued use of rags and cloth
made large-scale production impossible, until in 1800 Mathias Koops, a
Dutchman, patented a paper made from straw and wood. Three years later the
first functional paper-making machine was built at Frogmore, Kent, by the
engineer Bryan Donkin. The principle of his machine – pouring a suspension
of fibre and water onto a vibrating wire mesh conveyor belt to produce a
continuous sheet of paper – remains in use.

Pen and ink
Tools for inscription

Words in pictures, or pictographs, were drawn on cave walls with ink.
Pre-historic people used both black pigment made from charcoal and iron
oxide in shades between bright yellow and dark brown, mixed with animal fat,
to form viscous writing materials. The ancient Egyptian scribes worked with a
palate holding two 'ink cakes'. Black (from carbon) and red (from ochre) were
mixed with gum, dried, then rubbed with a wet brush made from rush stems
to make a liquid. In 2000 BC Chinese scribes used a durable ink made from
soot mixed with gum solution, writing on wooden strips using a bamboo pen.

In the 3rd century BC the Romans used the same method of writing as the
Egyptians, and also scratched text with a stylus, a sharp pointed tool, into a
wax-coated tablet. The advantage of writing on wax was that the tablet could
be easily smoothed over and reused.

Two new inks appeared in medieval Europe. One was prepared by combining iron salt and oak galls, producing a liquid that became dark brown with age. The other was made by suspending carbon from soot in a mixture of gum and water, but this needed constant stirring to stop it solidifying.

By the 13th century quill pens, made of sharpened goose feathers, were used throughout Europe. Such quills had been used by the Romans from about 500 BC, reappearing in Spain in the 7th century AD. Metal pens were made at least from the late 16th century but these were rare and largely ornamental. A steel nib was developed in 1803 by Bryan Donkin, but it proved less flexible than the quill and was corroded by the ink.

By the late 1820s nibbed pens were commercially available. In 1832 a John Joseph Parker became the first manufacturer to overcome successfully the difficulties of producing a pen that had its own reservoir of ink. But fountain pens did not catch on until, in 1884, the American inventor Lewis E Waterman developed the first practical version using the noncorrosive inks that were newly available.

Various designs of ball-point pens appeared after the first was patented in 1888 by the American inventor John L Loud. But they did not become commercially successful until 1938, when the Hungarian brothers Ladislao and Georg Biro produced a model using quick-drying ink. Launched in 1943, it sold by the million.

WRITING WITH 'LEAD'

A large oak tree in Borrowdale, Cumbria, is said to have been uprooted during a fierce storm in 1564, revealing a deposit of graphite, described then as 'black lead'. No one knows who first realised that this would make an ideal marking substance, but in 1565 the Swiss-German naturalist Conrad Gesner described a stylus of graphite encased in wood.

In 1858 the American Hyman Lipman had the idea of producing a pencil with a glued-in rubber eraser at one end. Four years later he sold the patent for his implement for $100,000.

The typewriter
Keyboard printing

Handwriting continued in offices until typewriters usurped calligraphy's monopoly. Henry Mill, the English inventor, had patented an artificial writing machine in 1714, but the earliest typewriter was not built until 1808. It was designed for the blind Italian Countess Carolina Fantoni to write her letters, but its mechanism remains unknown.

In 1866 two Americans, Christopher Latham Sholes and Carlos Glidden, produced a workable typewriter. Seven years later Sholes signed a contract with E. Remington and Sons, the New York gunmakers, who made the first mass-produced typewriters. One early buyer was Mark Twain, the first author to submit a typewritten manuscript, of *Life on the Mississippi*. The electric typewriter was invented in 1872 by American scientist Thomas Alva Edison and first sold by a Connecticut firm, Blickensderfer, in 1902.

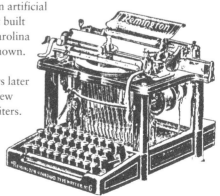

Print spreads the word

'He who first shortened the labour of copyists by device of *Movable Types* was disbanding hired armies, and cashiering most Kings and Senates, and creating a whole new democratic world: he had invented the art of printing.'

SARTOR RESARTUS, THOMAS CARLYLE, HISTORIAN AND ESSAYIST, 1833-4

The first prints were woodcuts, made by the Chinese from at least the 7th century AD. They cut spaces around an image on a flat wooden surface to make a raised 'negative' and then applied ink to it and transferred the image onto paper. This process of relief printing is thought to have evolved either from 'chops', ornately designed seals of Chinese characters used for stamping documents, or from the common practice of making inked rubbings from inscriptions. Like paper, relief printing spread west to Europe at a snail's pace. It was not until the 14th century that designs were printed on European textiles, and playing cards made from pieces of printed paper.

The development of typography – printing which used movable pieces of metal, each with a raised letter – made possible the mass production of letters, words and phrases on pages. Because the Chinese language consists of many thousands of characters and symbols, China was not the ideal place for movable type to evolve. The new process developed, instead, in medieval Europe.

First impressions

In about 1450, Johann Gutenberg perfected the art of printing in Mainz, Germany. He discovered a relatively soft alloy made of antimony, lead and tin. With this Gutenberg was able to cast letter shapes and found that they were durable enough to withstand repeated use. He also devised an ink to coat the metal, and adapted wine and paper presses to imprint ink on to a page. Among Gutenberg's first books were the Donatus *Latin Grammar* of about 1450, the 42-line Latin Bible (so called from the number of lines in each column), completed by 1455, and the *Catholicon*, an early encyclopedia of 1460. No more than 200 copies of each book were printed. The pages were decorated with hand-painted illuminations. However, printed books began to spread rapidly across Europe and by 1500, more than 9 million copies had been published.

Although the printing process remained basically unchanged for 400 years, different styles of lettering or 'typefaces' evolved. Gutenberg's first typeface imitated the handwriting styles of local scribes. Modern typefaces, which are designed on computers by creating letters on grids consisting of small squares called 'pixels', trace their origins back to 17th-century France. A committee of scholars was established by Louis XIV in 1692 to develop a new typeface based on 'scientific principles'. By dividing a page into more than 2,000 tiny squares, the scholars proceeded to design a precise typeface, Romain du Roi, in a similar manner.

From the 15th to the 20th century printers kept movable metal type in a 'typecase'. From here selected letters were made into words on a 'composing stick', adjustable to a column width by a screw or lever. Once the stick was full, the block of type was put on a tray known as a 'galley' – the name given to a printed page proof.

Lithography, named from the Greek words for 'stone writing', was invented in about 1796 by Aloys Senefelder, a Bavarian playwright, and used to print his dramas. In this technique a greasy crayon was used to draw on polished limestone. The stone was then wetted and, since grease repels water, the greasy areas remained dry. Greasy ink was then applied, which adhered only to the crayon grease. Paper laid on the stone then picked up the ink and created a reproduction of the original drawn image on paper.

Hot metal

The Linotype machine was patented in 1884 by Ottmar Mergenthaler, a German-born American inventor. It cast lines of type from molten metal, and was controlled by a typesetter operating a keyboard. Whole pages of type could be created this way, without having to arrange letters individually by hand. Two years later early models of the Linotype machine were used to print the *New York Tribune*.

In 1939 the American physicist Chester Carlson patented the first copying machine, which was operated by a process he called 'electrophotography'. It was later named 'xerography', from the Greek for 'dry writing'. Documents placed on a glass screen were copied by having their image focused on an electrically charged plate dusted with a negatively charged powder. The powder then stuck to the positive charge left by the image, which was transferred, with the help of heat, to a blank sheet of paper. The Haloid Company in New York welcomed the invention in 1947 and in 1949 the Xerox Copier Machine Model A was launched.

From clay to parchment
Knowledge bound in books

As the first civilisations developed they needed to record information so that it could be circulated and preserved. The Sumerian system of writing known as cuneiform evolved from simple pictographs around 3500 BC and was later used to inscribe royal annals, legal codes and epic stories on rectangular clay tablets. Particularly important ones were cased singly in clay 'envelopes'.

Papyrus, the main alternative to clay, was first used in ancient Egypt. Individual sheets were stuck together to form long scrolls, and rollers were attached at either end. As readers progressed through the text, they furled one roller while unfurling the other.

In China, narrow strips of wood or bamboo inscribed with text were lashed together with cords to create small works from about 1300 BC. Wax tablets bound into books appeared in the 8th century BC in Assyria. Each of the beeswax 'sheets', dyed yellow with an arsenic-based pigment known as orpiment, was contained in a frame of ivory or wood and inscribed with a bone or bronze stylus. Hinged along one side, the frames made a book that folded out like a fan.

Around the 1st century AD a Roman *codex*, mainly used for inscribing laws, appeared. Consisting of parchment leaves fastened together, this was the direct ancestor of the modern book.

Papyrus was made from reeds which grew only in certain places, and by the 4th century it had been superseded throughout Christendom by more widely available parchment. To create a book a large sheet of parchment or vellum (made from calfskin) was folded into a folio of two pages, a quarto of four or an octavo of eight pages. Grouped together into sets of leaves or quires, they were first cut, then sewn together with cords before being attached to a backing of leather.

THE PHILOLOGICAL SOCIETY WAS ESTABLISHED IN 1842 TO PREPARE THE FIRST ENGLISH DICTIONARY TO BE BASED ON HISTORICAL PRINCIPLES. AFTER SEVERAL DECADES OF RESEARCH THE DICTIONARY WAS COMPLETED IN 1928. FIVE YEARS LATER THE 13 VOLUME, 16,400 PAGE WORK WAS PUBLISHED UNDER ITS NEW TITLE, *THE OXFORD ENGLISH DICTIONARY*.

Encyclopedias and dictionaries
Reference collections

One of the earliest printed books was Johann Gutenberg's *Catholicon* of 1460, which continued a tradition of encyclopedias dating back to the ancient Greeks. In the 4th century BC the philosophers Aristotle and Plato had planned a cultural survey of society, incorporating philosophy and natural history. Fragments of their project, the first encyclopedia, survive in the work of Plato's nephew, the scholar Speusippos. By the

1st century AD encyclopedias had taken their modern form as vast anthologies of facts, such as the Roman author Pliny's *Historica Naturalis*.

During the Age of Enlightenment, in 18th-century Europe, the encyclopedia once again assumed a position of importance. Edited in 35 volumes by the French philosopher Denis Diderot between 1751 and 1780, the *Encyclopédie* offered a complete review of the arts and sciences of the day. It was a far more ambitious project than the publisher intended when he commissioned Diderot to translate the Englishman Ephraim Chambers's *Cyclopaedia*, the first edition of which appeared in 1728. The *Encyclopaedia Britannica*, first published in 1768, was created to bring some objectivity to Diderot's highly personal work.

Reference information was also compiled in ancient dictionaries. The Akkadians of Mesopotamia produced a list of words in the 7th century BC – the oldest known dictionary. But only in the early 18th century AD was a definitive work on the English language completed. In 1707 Humphrey Wanley, a member of the Royal Society, requested a work to 'fix' the English language. Samuel Johnson, the English writer, responded, and with six assistants recorded 43,500 words and 118,000 illustrative quotations by 1755.

BOOKS TO BORROW

* Founded in the 3rd century BC at Alexandria in Egypt, the largest Greek library was reputed to hold as many as 700,000 scrolls.
* In the 17th century the Stationer's Company, the government's publishers, resolved to give the Bodleian Library in Oxford a copy of every book published in Britain. The British Museum Library, now the British Library, was awarded the same privilege in 1753.
* Subscription libraries flourished in 18th and 19th-century Britain. Free public lending libraries were established by Act of Parliament in 1850.

Popular reading

Books for all

Popular paperbacks have their roots in 16th-century Italy, where small, portable and relatively inexpensive texts were first published in Venice by Aldus Manutius in 1501. To make his scholarly titles accessible to a wider audience, Manutius printed runs of 1,000, instead of the usual 250 copies, and used an italic typeface to fit more words onto each page. Since his methods were widely imitated he devised one of the first publisher's logos, a dolphin and anchor, to protect his editions against unauthorised copying.

The novel, the only literary genre to appear after the invention of mechanical printing, revolutionised reading habits. Prose fiction had been available since the 1650s, but it was *Robinson Crusoe*, Daniel Defoe's novel of 1719, that brought secular books to a mass audience.

By the 1820s cloth cases around stiff covers began to replace leather, so that in the 1850s George Routledge's Railway Library of novels could be sold for just one shilling (five pence). But the true paperback was not created until 1935 when Allen Lane launched his Penguin series. As a travelling salesman weighed down with casebound books, Lane had decided that the future lay in paper covers.

Reporting events
The birth of newspapers

THE FIRST SUNDAY PAPER, LAUNCHED IN 1779, WAS ELIZABETH JOHNSON'S *BRITISH GAZETTE AND SUNDAY MONITOR*, WHICH CONTAINED A SUMMARY OF THE WEEK'S NEWS AND A RELIGIOUS COLUMN. 'IT KILLS A FEW HOURS OF THIS DULL MORNING AND PLEADS AN EXCUSE FOR A PREFERENCE OF A COFFEE-HOUSE BOX TO A CHAPEL-PEW,' QUIPPED WILLIAM DAVIES IN HIS SATIRE OF 1786, *NEWS THE MALADY*.

In the public meeting places or *fora* of ancient Rome, in about 60 BC, the first news reports were posted on walls to keep the public informed. A government gazette, the *Acta Diurna*, or *Daily Events*, featured official announcements together with reports of battles, gladiatorial games and astrological omens.

A system of mass news reportage was also developed in China during the Han dynasty (202 BC to AD 220). Summaries of events written by postmasters in distant parts of the empire would be delivered back to court. The *Pao*, or *Peking Gazette*, a sheet distributed to civil servants, appeared during the Tang dynasty (AD 618 to 906) and provided the latest court news.

'Broadsheets' were originally single pages published in late 15th-century Europe by the Church and State. They mixed propaganda with sensationalised reports of such events as floods and mystical visions, and were illustrated with woodcuts to aid the illiterate.

BREAKING NEWS

A short pamphlet of 1513 entitled *The True Encounter* is the first known published account of a contemporary event by an eyewitness. Describing the Battle of Flodden, in which the English army defeated the troops of James IV of Scotland, it was written and printed by an Englishman, Richard Faques.

In the 17th century single-page news sheets printed with 'relations' of events were read by an educated elite some time after the events had taken place. There were also 'diurnalls' (weekly reports) in book form, political 'mercuries' (named after the Roman messenger of the gods) and government news 'intelligencers'.

Dutch *corantos*, 'currents of news', appeared in the 1620s. They provided news of the week's political events in Europe, printed for the first time on folded sheets of paper, the form used by today's newspapers. The first daily was *The Daily Courant*, which reported on international affairs to a small London readership from 1702. *The Evening Post*, first printed in 1706, was aimed at a wider audience and was published three times a week to coincide with the mail coaches leaving London.

'News' writers were plying their trade by the 1600s in Japan. They compiled single-page gossip sheets, containing information collected by traders who travelled around the country. News gatherers in 18th-century London congregated around the Royal Exchange and St Paul's Cathedral. After gleaning business and political news from men of affairs, they would rush to sell information to publishers. Full-time reporters were first employed by newspapers in the 19th century.

The popular press
Mass circulation

Headlines to catch readers' attention evolved from the 17th century. By the 1770s the events of the American War of Independence merited such punchy titles as 'Detroit is Taken'. In Britain taxes on paper meant editors avoided such an extravagant use of space. Only after taxes were abolished in 1855 were banner headlines truly exploited.

The Times, the oldest surviving title in Britain, was launched by a bankrupt coal merchant, John Walter, to pay off his debts. It was originally published as the *Daily Universal Register* in 1785. Appealing to an educated readership, it prospered as *The Times*, the enduring name it was given in 1788. It was also known as 'The Thunderer', a nickname it lived up to when it published an article on the Reform Bill in 1831, urging people to 'thunder for reform'.

The first tabloid – with half-size pages – was produced by the British editor Alfred Harmsworth on January 1, 1900, as a special millennium issue of New York's *The World*. He relaunched the *Daily Mirror*, his ailing newspaper for 'gentlewomen', as Britain's first tabloid in 1904, with immediate success.

Periodic papers
The rise of the magazine

The *Gentleman's Magazine*, published from 1731, was one of the first magazines to use the word, derived from the Arabic *makhzan*, meaning 'storehouse'. Periodicals had made their appearance in the previous century. The edifying *Monthly Discussions* was the first, published by the German theologian Johann Rist in 1663. Nine years later in France a more frivolous type was launched with *Le Mercure Galant*, containing poetry and court gossip.

In England the similarly lightweight *Athenian Gazette* (retitled the *Athenian Mercury* after one issue) began publication in 1690. Made anxious by an extramarital affair he was conducting, its publisher John Dunton had the idea of the 'problem page' in 1691. The advisers, the original 'agony aunts', were in fact men. Rising literacy levels among women led Dunton to launch *The Ladies' Mercury*, the first magazine for women, in 1693.

Literary magazines appeared in the 18th century. Among the first was the *Museum* of 1746, which mainly printed book reviews. In the 19th century magazines were used as forums for serialised novels. Several works of Charles Dickens were printed in magazines that he himself edited, including *The Old Curiosity Shop* which appeared in 1840–41 in *Master Humphrey's Clock*.

JOURNALISTIC MILESTONES

✻ In 1621 the printer Nathaniel Butter published an English newspaper in London. It had no fixed title and appeared, more or less, weekly.

✻ Mary de la Rivière became the first female newspaper editor, of England's *The Examiner* in 1711.

✻ *Punch*, a magazine dedicated to satirising the British way of life, was launched in 1841.

✻ *Good Housekeeping*, which tested consumer goods, was published in 1885 in the United States.

✻ In 1888 Britain's *Financial Times* appeared, printed on pink paper to distinguish itself from its rivals.

The birth of broadcasting
Mastering the airwaves

Radio waves, theorised the Cambridge physicist James Clerk Maxwell in 1864, could not only be artificially created but would travel at the speed of light. Maxwell was correct, and in about 1888 Heinrich Hertz, a German physics professor, transmitted such 'radiation' (the source of the word 'radio') from a spark generated between two metal balls.

Practical radio was made possible by the Irish-Italian Guglielmo Marconi. In 1894 he used Hertz's spark generator to send a telegraph message across a room in the family home near Bologna. Marconi soon discovered that by using aerials and connecting an earth to both transmitter and receiver radio range could be increased. In 1895 he transmitted a message to his brother Alfonso, unseen on the opposite side of a hill. The first transatlantic radio signal was sent from Cornwall to Newfoundland in 1901.

Radio buffs also made progress with homemade sets that relied on the detection qualities of crystals such as silicon, discovered in the early 1900s. These were incorporated into 'cat's whiskers', receivers named after the thin wire that was used to connect the crystal with an electric circuit.

In 1922 the Operadio, portable but weighing 10kg (22lb), was invented by the American J McWilliams Stone. Radios were miniaturised after the transistor was invented in 1947 by John Bardeen, Walter Brattain and William Shockley, scientists for Bell Telephone Laboratories in the USA. The Sony TR-55 was launched in 1955, quickly followed by other transistor radios.

Radio entertainment
Popular programmes

Beginning with 'O Holy Night', which he played on the violin, the Canadian-born Reginald Aubrey Fessenden broadcast carols on Christmas Eve 1906 from Brant Rock, Massachusetts. His tunes were gladly received by sailors 8km (5 miles) out at sea. But radio entertainment was slow to develop. The American Lee de Forest transmitted live opera from New York in 1910. Later, in 1917, the German army broadcast radio entertainment for its troops. The first regular radio station was KDKA, founded in Pittsburgh, USA, in 1920.

Once the British Broadcasting Company was established in 1922 radio finally became available to the British public. The inaugural BBC programme, a news broadcast, was transmitted in November. The Light Programme (now Radio 2) was launched in 1945, followed by the Third Programme in 1946. Pop music first came on air with Radio 1 in 1967: the station's disc jockey Tony Blackburn inaugurated the station by playing 'Flowers in the Rain' by The Move.

CHAPTER 6

TRANSPORT AND TRAVEL

In 1895, the first ever motor race was run from Paris to Bordeaux and back again. It marked the beginning of a new era, when people would try to travel as fast as possible, for the sake of speed alone. That first race took the winning car 49 hours to complete the course, travelling at an average speed of 24km/h (15mph). Just eight years later competitors in the 1903 Paris to Madrid race were topping 160km/h (100mph).

When early man created the first form of transport – the boat – his desire was simply to carry himself safely across water: the first humans walked out of Africa, crossing seas and other waterways in their journey to colonise the rest of the globe. On land, domesticated animals were used to carry people and goods until the invention of the wheel – the first wheeled wagons were made in Sumeria in around 3500 BC.

Speedy land transport wasn't really possible until the 18th century, when roads were improved during the Industrial Revolution. When trains were introduced in the 1830s people were able to travel faster than ever before. But it was the development of the internal combustion engine at the end of the 19th century – which lead to the invention of the car and the airplane – that would change human transport for ever.

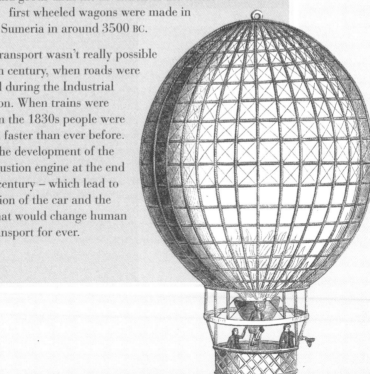

The rise of the bicycle
Pedal power

The bicycle is a simple machine but its design was not easy to perfect. In fact the development of the 'iron horse', which enabled thousands of ordinary men and women to escape from the grime of Victorian towns and discover the joys of the countryside, occupied most of the 19th century.

The first person to realise that it was possible to construct a balanced two-wheeled vehicle was Baron Karl Drais von Sauerbronn, a German engineer who showed his wooden Draisine in 1817 in Paris. The following year, English coachmaker Dennis Johnson copied the design to produce an iron model.

More than two decades later, a Scottish blacksmith called Kirkpatrick Macmillan developed swinging pedals that operated cranks driving the vehicle's rear wheel. But his design failed to catch on and the next breakthrough did not occur until 1861, when the French coachbuilder Pierre Michaux and his son Ernest produced a *vélocipède*, named from the Latin for 'swift-footed'. The large front wheel had revolving pedals attached to its hub, while a simple brake acted as a control on the rear wheel.

With this principle of construction, the bigger the front wheel the further the velocipede could travel for each turn of the pedals. The concept was taken to an extreme by James Starley, foreman at the Sewing Machine Co. of Coventry. His 1870 model had the rider perched precariously on a wheel measuring up to 1.5m (5ft) across, with a diminutive rear wheel trailing behind. Nicknamed the penny-farthing from the sizes of the wheels, it created the first cycling boom. Tricycle versions were made for those of a nervous disposition and for women, who found it hard to manage pedalling with their long and voluminous skirts.

THE TANDEM FIRST TOOK TO THE ROAD IN 1869, WHILE A TRICYCLE FOR TWO AND A 'SOCIABLE' FOR THREE WERE PRODUCED IN 1883. BY THIS TIME THE CYCLISTS' TOURING CLUB OF GREAT BRITAIN, ESTABLISHED IN 1878, HAD SOME 10,000 MEMBERS. 'THE WORLD REVOLVES AROUND THE BICYCLE,' ENTHUSED NEW RECRUITS TO THE CRAZE.

SAFER CYCLING

'The horse that never says neigh', as the penny-farthing was dubbed in the USA, was used for outings, racing and a round-the-world trip. The adventurer Thomas Stevens set off on his machine from California in April 1884 and covered 21,725km (13,500 miles) before returning in December 1886. By this time, the penny-farthing's popularity was being challenged by the Rover 'safety' bicycle, which was first produced in 1885 by John Kemp Starley, James Starley's nephew. The Rover laid the foundations for the modern bicycle: two wheels of equal diameter, with the rear one driven by a chain that was turned by pedals set between the wheels, and a diamond-shaped steel frame.

The final element was the pneumatic tyre, an instant hit on its launch in 1888 when it was advertised with the boast 'vibration impossible'. In 1887 an Irish vet, John Boyd Dunlop, was watching his son bounce uncomfortably on a tricycle and he hit upon the idea of fashioning a cushion for the rims of the wheels from a garden hose, filled with water. Later, he improved upon the design by using rubber inflated with air. The pneumatic tyre was a great success but the idea was not new – the civil engineer Robert Thompson had first patented it in 1845. With the application of gears, developed in France in 1889 to minimise the effort needed to pedal over varied gradients and terrains, 'bone shakers' were things of the past.

The bicycle also played a part in emancipating women's clothing. In the late 19th century some female cyclists adopted loose trousers gathered at the ankle, named 'bloomers' after their advocate Amelia Jenks Bloomer of New York. Cycle racing began in 1868, with the Englishman James Moore winning a track race in Paris. He won the first road race, a 134km (83 mile) trek between Paris and Rouen, the following year.

Motorcycles
Born to be wild

The bicycle is the most fuel-efficient mode of transport yet invented, but the cyclist cannot escape the necessity of exerting muscle power. In 1868, just a few years after they had developed their rotary-pedal velocipede, Pierre and Ernest Michaux attempted to resolve this drawback by attaching a small steam engine to its rear.

However, the first bicycle powered by a petrol engine was built in 1885 by the German engineer Gottlieb Daimler, whose main attention at the time was focused on the invention of the motor car. The motorcycle acquired its classic layout, an engine low in the frame between the wheels and controlled by twist grips on the handlebars, at the instigation of two Paris journalists, the brothers Eugene and Michel Werner. Their *motocyclette* was exhibited at the 1897 Paris Salon.

> ## MODS AND ROCKERS
>
> In the early 1950s, the NSU company in Germany produced mopeds, small-engined machines started by pedalling. Motor scooters, which had more powerful engines and 'skirts' around the front to provide protection from the weather, had first appeared in Italy in 1946 when the Piaggio company introduced its Vespa. The Lambretta was launched by a rival company in 1947. By the 1960s 'mods', who favoured scooters, and 'rockers', who preferred motorcycles, had become distinct, and often violently opposed, youth groups in Britain.

Development of the motorcycle was rapid, and was accelerated considerably by the establishment of such races as the Tourist Trophy (TT), first held on public roads on the Isle of Man in 1907. At the first motorcycle race, a 152km (94 mile) course from Paris to Nantes and back, organised by the Automobile Club de France in 1896, all but one of the competitors rode three-wheeled models. But the two-wheeled variant proved best for manoeuvrability. The invention of the side car in 1903 by a W G Graham killed off the trikes, which were often built for more than one person, but in recent years they have made a comeback.

Cabs and taxis
Transport for hire

London's black-cab drivers, who must pass the rigorous test known as The Knowledge and become licensed before they can ply their trade on the street, continue a tradition that was established in 1662. Needing funds for road improvements, Charles II introduced licensing in that year under an Act of Parliament which included regulations for hackney coachmen.

The light, two-wheeled cabriolet, named from the French for a 'playful leap' and introduced in 1823, gave rise to the generic term 'cab'. Other specialist hire coaches included the hansom, a two-seater with an enclosed carriage patented in 1834 by the Leicester architect Joseph Hansom. The brougham, built for Lord Brougham by the London coachbuilders Robinson of Mount Street, was introduced in 1839.

In London the motorised taxi only caught on after 1907, when the reliable two-cylinder Renault was imported from France. As with horse-drawn models, designs for motorised cabs proliferated. The introduction of a mass-produced model, the FX3, in 1946, gave rise to the now familiar black cab.

THE BABY CARRIAGE WAS THE FORERUNNER OF THE PRAM, A SHORTENED FORM OF 'PERAMBULATOR', FROM THE LATIN FOR 'WALK ABOUT'. ENGLAND'S EARLIEST KNOWN EXAMPLE WAS SHAPED LIKE A SCALLOP SHELL AND FEATURED A CANOPY ROOF AND CARRIAGE LAMPS. DEIGNED TO BE PULLED BY A DOG, THE WOODEN VEHICLE WAS CREATED FOR THE 3RD DUKE OF DEVONSHIRE BY THE ARCHITECT WILLIAM KENT IN 1733.

Coaches and buses
People on the move

Journeys were exhausting for the horses pulling the stagecoaches that began to link Britain's major towns in the 17th century. Travelling between infrequent stops or stages, at which they were replaced, the teams of horses that plied the rutted highways might cover no more than 25km (15 miles) in a day.

In the early 19th century, passengers on the newly established fast mail coaches could reach Bath from London in 15 hours instead of three days, but road travel was still, in the words of the novelist Charles Dickens, 'a very serious penance'. However, scheduled public transport was now at least reliable and increasingly affordable.

PUBLIC PROGRESS

The early experiments in urban public transport, such as the scheduled service in Paris started by the French mathematician and philosopher Blaise Pascal in 1662, were no more than briefly fashionable with the aristocrats on whom their success depended. By the 19th century, however, the expansion of towns and cities meant that many more people had to travel long distances to reach their places of work.

The linking of the word 'omnibus', Latin for 'everything' or 'everybody', with a public transport vehicle began in Nantes in Brittany in about 1823. Stanislas Baudry began operating his 16-seat horse-drawn vehicle to a regular timetable, taking customers from the centre of town to the bathhouse he ran in the suburbs. If space allowed, Baudry would carry other passengers, and he used the word 'omnibus' to emphasise his service's universal appeal. The bathhouse soon shut down but the omnibus service remained hugely successful.

In London the first omnibus route was launched in July 1829 by George Shillibeer, who had worked as a coachbuilder in France. The vehicles, which ran from Paddington Green to the Bank of England, had enough seats for 20 passengers. Staff were uniformed and courteous, and the passengers were offered free newspapers and magazines to help them to pass the journey. The vehicles were called 'Shillibeers' because 'omnibus' was considered a vulgar word. When rival operators painted his name on the sides of their vehicles, Shillibeer began calling his own 'Shillibeer's Original Omnibuses', and the word became part of polite vocabulary. Double-decker horse-drawn buses were introduced in 1847. They became widespread when the Great Exhibition of 1851 led to a vast increase in passengers in London.

Early attempts at motorbuses were failures. Steam coaches, which were introduced in the 1830s, were too heavy and damaged road surfaces. They faced opposition from the operators of turnpike roads, as well as heavy charges. The first petrol-driven buses, which operated in London from 1899, proved unreliable. However, there was a breakthrough when the London General Omnibus Company introduced its dependable B-type in 1910. In 1913 B-type buses lost only 0.02 per cent of their scheduled time to breakdowns.

The caravan

Home on the road

The circus proprietor Antoine Franconi owned one of the first caravans, which appeared in France in the 1830s. His elaborately decorated *voiture nomade* contained a built-in kitchen, dining room and bedroom. Such vehicles were also known in Britain. In his novel *The Old Curiosity Shop*, which was published in 1840, Charles Dickens describes in detail the fictional caravan of Mrs Jarley, the owner of a travelling waxworks.

The concept of caravans may have been new in the 19th century, but the word caravan, from the Persian *karwan*, 'company of travellers', was not. The term had been used to refer to a basic stagewagon for 200 years. However, decorated Gypsy caravans did not become familiar sights on British roads until the 1870s. The explorer Sir Samuel Baker pioneered the idea of the caravan pleasure tour when he and his wife purchased a Gypsy caravan and toured Cyprus for six months in 1879, pulled by a team of oxen.

DRIVING FORCES

✤ Horse-drawn wagons running on rails, called trams after the rail carts used in mines, appeared in New York in 1832, in Paris in 1855, and in Birkenhead in 1860. Electric trams were used in Germany from 1884 and in Leeds from 1891.

✤ The Paris Automobile Club hired out the first rental cars in 1896. Britain's first hire service, begun by James Edward Tuke of Harrogate and Bradford, Yorkshire, opened for business in the same year.

✤ The first motorised hearse was used in Buffalo, New York, in 1900. Britain's first was a Daimler used in 1901 to transport the body of William Drakenford, a Daimler Motor Car Company employee.

The car is born

'Glorious, stirring sight!' murmured Toad ...
'The poetry of motion, the real way to travel ...
O bliss! O poop poop! O my! O my!'

THE WIND IN THE WILLOWS, KENNETH GRAHAME, 1908

It looked like a rickety tricycle, but the contraption tested out by the German engineer Karl Benz in the grounds of his Mannheim workshop in the spring of 1885 was destined to change the face of the world. The following summer, the motor car covered about 1km (⅔mile) at a speed of 15km/h (about 9mph) in a public demonstration. This epoch-making event was reported in a local newspaper under the heading 'Miscellaneous'.

While Benz was busy in Mannheim, Gottlieb Daimler, another German engineer, was assembling his first motor car, a four-wheeled carriage. After many false starts, the time was ripe for the birth of the car.

Steam engines had been used to drive road vehicles since the late 1700s, but they were heavy and inefficient. Motor cars needed a new prime mover. It was found in the internal-combustion engine, in which fuel is burned inside the engine rather than in an external furnace, such as in a steam engine.

The first practical version of such an engine, running on coal gas, was developed in 1860 by Jean Joseph Etienne Lenoir, a Belgian engineer working in Paris. Lenoir ran a vehicle from Paris to Joinville-le-Pont, a journey of some 9.5km (6 miles), which took about 3 hours.

Effective engines

In 1862 the French scientist Alphonse Beau de Rochas proposed a more efficient system for the internal-combustion engine known as the four-stroke cycle, but his idea came to nothing. More than a decade later the German engineer Nikolaus August Otto effectively reinvented the engine with his four-stroke gas-powered version, manufactured in 1876. Daimler and Benz used Otto's four-stroke cycle to make engines that could run on petrol and were about nine times more powerful than Lenoir's. Although both produced their cars in 1885, Benz was the first to offer his vehicles for sale, in 1887.

From its early days in Germany, the motor car was destined to progress fastest in France. In 1895 a car designed by René Panhard and Emile Levassor, with features providing the pattern for the modern car, including a wheel on each corner, the engine at the front and a pedal clutch, won the first motor race. It took just under 49 hours to complete the course from Paris to Bordeaux and back, at an average speed of 24km/h (15mph).

He finished 6 hours ahead of the second car, a Peugeot which sported the first pneumatic tyres to be used on a car. The Peugeot might have won the race had its driver not exhausted his stock of 22 inner tubes for puncture repairs. Barely a decade after its invention the car had proved itself much more than a toy.

Speed limits

In Britain, motor-car development was held back by the Highways and Locomotives Act, introduced in 1865 to control heavy steam-powered traction engines. Mechanical vehicles were restricted to a maximum of 4mph (6.5km/h) and had to be preceded by a person carrying a red flag. The first British road motorist was a Henry Hewetson of Catford, who imported a Benz from Germany in November 1894. He hired a young man to cycle ahead of the car and warn him if he spotted a policeman.

Despite the restrictions on driving, some people accurately foresaw a profitable future for the motor car. One of them was the entrepreneur Frederick Simms, who bought the patents to Daimler's engines in 1893. When the 'red flag' Act was amended in 1896 to allow 'light locomotives' to be driven on the roads, the Daimler Motor Car Company became the first car manufacturer in Britain. It completed its first car in 1897.

Building these early cars took skilled craftsmen months but in 1913 the American car manufacturer Henry Ford installed a conveyor-belt assembly line at his factory: a single car could now be put together in about 90 minutes. In contrast to other cars of the time, which were far beyond the means of the average family, Ford's Model T, introduced in 1909, was designed as a 'car for the multitude' and was priced accordingly. This was the car that launched the era of popular motoring and in 1920 it became the first car to sell a million.

Mass production reached Europe after the Second World War. The first British model to sell a million was the Morris Minor, launched in 1949.

The early motorists were dedicated enthusiasts, and needed to be since cars were both uncomfortable and unreliable. As more people acquired cars, they needed to be taught how to use them. The first dedicated driving school was opened in Birkenhead by a William Lea in 1901. Driving tests and licences became compulsory in Paris in 1893. In Britain licences were required after 1903, but were issued annually on request regardless of competence. Driving tests were introduced in 1935.

Highways and byways
How roads were made

Throughout history rulers have needed good roads to ensure control of their domain. The Romans built roads throughout their empire, including Britain, from the 1st century AD. After the fall of the empire, roads were left to decline and Britons went back to using roads that were little more than dirt tracks.

Often these roads were so steeply cambered that carriages were in danger of toppling over. Only a few had a paved causeway, or 'causey', alongside for pedestrians and pack animals.

At the other extreme were the 'hollow ways', which are still found in some parts of England, notably Devon and Cornwall. Streams or rainwater could be diverted along these concave roads to wash the mud and rubble to the lowest point, from where it could be cleared away easily. Gradually these roads sank lower and lower.

The first motorway was the 10km (6¼ mile) Avus Autobahn, on the outskirts of Berlin, opened in 1921. Its instigator was the racing enthusiast Karl Friedrich Fritsch, who provided a loop at each end so that the road could double as a racing track.

From the 18th century, some British streets were 'paved' with naturally rounded cobblestones from beaches – some of these still survive in Rye, Sussex. Blocks cut from hunks of hard stone, usually granite, were also used and gaps between were filled with earth. This quickly became compacted so water could not penetrate, producing a reliable, if bumpy, surface.

Pierre Trésaguet, a French road engineer, created a more stable road structure in the 1760s by having three layers of successively smaller stones on a cambered earth base. Similar methods were applied by the Scottish civil engineer Thomas Telford in the early 19th century. In his roads, grit produced by wheels rolling over a top layer of broken stone filled any gaps, making the surface watertight.

Telford's methods were simplified in the 1820s by his compatriot and fellow engineer John McAdam, who dispensed with foundations. So long as it was waterproof, he asserted, a single layer of graded stone laid on the earth would suffice. McAdam told his workmen to select only stones small enough to fit into their mouths. One day he came across a stretch covered with larger stones. When the workman was admonished, so the story goes, he grinned, revealing a huge toothless mouth. After that, the surveyors were given a gauge and a balance for checking the sizes of stones.

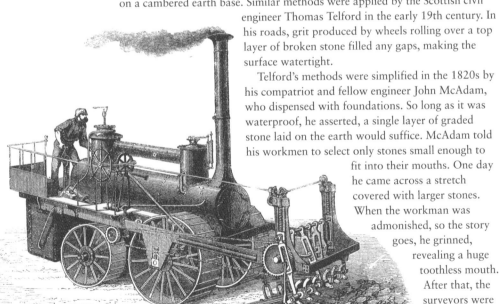

But the demands of the motor car called for a more impervious surface, not least to suppress the clouds of dust created by the suction of the tyres. The solution was to seal the surface with tar. Spread hot then compacted with a roller and top-dressed with gravel, 'tarmacadam', from the word 'tar' and McAdam, or 'tarmac' makes a smooth surface. A section of the London to Nottingham road was the first to benefit from this material, in 1845.

Road rules
Codes, signs and signals

Regulation of traffic did not begin with the invention of the motor car. As early as 45 BC Julius Caesar limited the number of wheeled vehicles allowed to enter Rome – exceptions were granted for those 'bringing materials necessary for building temples to the gods or public works' or those carrying priests or vestal virgins.

The Romans also introduced the custom of keeping to the left, a habit reinforced in medieval times when riders throughout Europe passed oncoming strangers sword arm to sword arm. An increase in horse traffic at the end of the 18th century meant that the convention gained strength but it was not enshrined in British legislation until 1835.

Until the Revolution of 1789, French carriages had habitually kept to the left, which forced pedestrians onto the right side of the road. However, when they were faced with oncoming crowds of hostile *sans-culottes*, or republicans, the aristocrats wisely moved over. Maximilien Robespierre, the revolutionary leader, ordered all Paris traffic to drive on the right in 1791. The following year the first 'keep right' law in the USA was applied to the Pennsylvania turnpike, after visits by Marie Joseph La Fayette, the French soldier and liberal reformer.

Roman milestones were the earliest incarnations of road signs. The prototype of the modern traffic sign was erected near Lausanne in Switzerland in 1790 to warn of a steep hill. In Britain the Bicycle Union placed the first road signs on dangerous hills, in 1879. Throughout Europe a plethora of signs followed until 1903, when the French pioneered the concept of nationally standardised symbols.

The first island at the centre of a road was built in Liverpool in 1862. In 1864 a Colonel Pierpoint had one sited in London's St James's Street so that he could reach his gentlemen's club safely. But this did not prevent him from being knocked down by a cab at that very crossing. It is said that the accident happened as he stood in the road admiring his handiwork.

UNDER CONTROL

✻ One-way traffic systems were used in some of London's narrow lanes in 1617 to regulate 'the disorder and rude behaviour of Carmen, Draymen and others using Cartes'.

✻ Speed limits of 4mph (6km/h) in the country and about half that in towns and villages were introduced in Britain in 1865 to control vehicles powered by steam. In 1896, after the motor car had become established, it was raised to 12mph (19km/h).

✻ Gas-lit traffic lights, using the colours red for 'stop' and green for 'caution', were placed near Parliament Square, London, in 1868. The three-colour system of red, amber and green was introduced into New York in 1918, and London in 1926.

FOR SAFETY—IN A CLASS BY ITSELF

Finding the way

'So geographers, in Afric-maps,
With savage-pictures fill their gaps;
And o'er unhabitable downs
Place elephants for want of towns.'

ON POETRY, JONATHAN SWIFT, 1733

The world's oldest surviving map was drawn on a small clay tablet in 2300 BC, discovered at Yorgan Tepe in modern-day Iraq. It shows an area bounded by hills and divided by a waterway, and names the owner of an area of land. Marked on its edges are north, south, east and west.

From the same area is the first known attempt to map the world, in around 600 BC. At the map's centre is Babylon and the River Euphrates; a ring of ocean forms an outer boundary. More practical is a map from ancient Egypt, a papyrus of 1150 BC showing the layout of gold mines near the Red Sea coast and how to get there.

By 600 BC Phoenician explorers were using the positions of the brightest stars to guide them on their journeys and the ancient Greeks produced maps of the constellations. Using their knowledge, observations and imagination, the Greeks also speculated about the nature of the Earth. By the middle of the 4th century BC they knew the Earth was round, and had devised the lines of latitude forming parallel horizontal bands north and south of the Equator.

Mapping the world

Lines of longitude, the meridians that run vertically from pole to pole, were added piecemeal until in AD 100 the geographer Marinus of Tyre developed a grid with regularly spaced intervals in which locations could be described mathematically by means of the grid's coordinates.

Building on Marinus's work, astronomer and geographer Ptolemy of Alexandria produced the first great world map in about 150. Ptolemy greatly influenced the development of geography, and more than 1,000 years later navigators were only just beginning to correct his calculations.

In the 13th century sailors developed the portolan, a form of navigational chart which recorded the distances and bearings between landmarks, sea depths and tidal variations. When the Genoese navigator Christopher Columbus set off on his first voyage across the Atlantic he was convinced he would reach Asia. Instead, in 1492, he landed in the Bahamas and revealed the existence of the Americas to Europeans. Thanks to his explorations and those of other navigators, maps became increasingly accurate in the 16th century. The foremost map-maker of the era was the Flemish Gerardus Mercator, who devised a method of displaying the curved surface of the Earth on a flat map. This enabled navigators to plot bearings as straight lines.

The introduction of the telescope in the early 1600s made it possible to accurately chart the positions of stars and other celestial bodies. This led to the establishment of the first modern observatories.

The Royal Observatory at Greenwich, was built in 1675 to record the positions of all stars. Sailors then calculated their longitude by observing the motion of the Moon against the fixed background of stars. But this method was not sufficiently accurate.

Technical developments

Latitude had long been estimated by observing the height or elevation of the stars above the horizon. The problem of estimating longitude while on the move was solved in 1761 by English clockmaker John Harrison. He devised a chronometer which would keep precise time at sea. By working out local time from the elevation of the Sun, and comparing it with the chronometer reading, which was set to Greenwich time, sailors could calculate how far east or west they were of the observatory.

The Sun's elevation could be measured with the reflecting quadrant, invented by the English astronomer John Hadley in 1730, and with the sextant, introduced by Captain John Campbell, after 1757. The Greenwich Meridian was accepted internationally as the prime meridian, zero degrees of longitude, in 1884.

The 18th century saw the first truly accurate maps created using triangulation, in which distances are calculated by measuring angles to a distant point from each end of a baseline. The French cartographer César François Cassini de Thury produced the first, mapping France between 1744 and 1783 at a scale of about 2.5cm to 1.2km (1in to ¼ mile).

The birth of locomotion
Full steam ahead

Grooved stone pathways to guide vehicles were used by the Babylonians in around 2000 BC, but it was the wagonways of the medieval mines in Europe that led directly to the railways. In the 1st century AD the inventor and mathematician Hero of Alexandria had experimented with steam, the motive power of the railway revolution. He played with a wind ball, or aeolipile, in which a hollow sphere connected by pipes to a sealed cauldron of heated water rotated as steam escaped.

Only in 1698 was steam used for a practical purpose. The English engineer Thomas Savery invented a pump that used vacuum power produced by condensation of the steam. But it was Thomas Newcomen, the Devonshire ironmonger, who built the first practical steam engine, in 1712. Like Savery's pump, it was used for draining water from mines. In the 1760s the Scottish instrument-maker James Watt improved the power and efficiency of the steam engine by introducing a separate condenser.

By the 1780s both British and French pioneers were experimenting with using steam to drive heavy-wheeled road engines. A breakthrough came in 1804 when Cornish engineer Richard Trevithick developed a high-pressure engine compact enough to operate on a tramway. After a few days the rails on which the locomotive was run cracked under its weight and the engine was converted into a pump.

PULLING POWER

Unable to obtain financial backing for his engines Trevithick left Britain to work in the silver mines of Peru. The challenge was then taken up by William Hedley of Newcastle upon Tyne. In 1813 he built the *Puffing Billy*, a locomotive christened for its exhalations of steam. It was George Stephenson, a mine mechanic, who was most tenacious in pursuing the development of the locomotive. In 1825 his *Locomotion* was used to pull the first engine-driven freight and passenger service along the 16km (10 miles) of the Stockton & Darlington Railway. Stephenson's son Robert built the *Rocket*, the engine that won the 1829 trials to decide the form of motive power for the Liverpool to Manchester line.

The German engineer Werner von Siemens designed the first successful electric train, drawn by a locomotive picking up current from a live rail, in Berlin in 1879. Four years later electric trains were introduced in Britain, in Brighton and Northern Ireland.

ON THE RIGHT TRACK

✤ Sleeping cars were introduced in the USA in 1836. Passengers had to supply their own bedding.

✤ The American George Pullman leased his first carriage in 1859. The drawing-room-cum-sleeper Pullman cars were the inspiration behind the luxurious Orient Express, established in 1883 by the Belgian entrepreneur Georges Nagelmackers.

✤ The American Union Pacific and Central Pacific Railroads completed the first transcontinental railway, running from coast to coast, in 1869.

✤ One of the earliest films set around trains or a station was *The General*, a 1926 silent comedy based on an incident in the American Civil War. Buster Keaton played a Confederate train driver chasing his train after it had been stolen by Union soldiers.

Diesel power, which did not involve the huge costs of electrification, made its debut in 1912, in a locomotive built by the Swiss firm Sulzer. However, credit for the technology lies with the German engineer Rudolf Diesel, who perfected the engine in 1897.

Passenger trains
Tickets and timetables

As train travel developed and started to accommodate passengers, it quickly became a matter of class. Panelled coaches, which had well-padded seats, allowed first-class passengers to relax in comfort. Second-class coaches were simpler in style but usually had seats, while passengers in third-class open wagons were showered with soot spewed out by the engine.

The first railway station opened at Liverpool Road, Manchester, in 1830, complete with a waiting room and separate booking halls for first and second-class passengers. A booking clerk named Thomas Edmondson, who worked on the Newcastle & Carlisle Railway, invented preprinted tickets in 1837 to save him writing out the same details for each traveller. For passengers unable to read, some railways issued tickets with pictures that represented major towns: a sack of cotton for Manchester, a fleece for Leeds, after their main industries.

As more rail services were set up, timetables became necessary. The first was published in 1838 by the London & Birmingham Railway. In 1839 George Bradshaw, a Manchester printer, produced the first timetable of all services. The growth of timetabled railway operations led to standard time being adopted throughout the UK. However, they gave 'no guarantee of punctuality', as the South Yorkshire Railway admitted in 1851.

> THE METROPOLITAN RAILWAY, A 6KM (3¾ MILE) LONDON LINE WHICH LINKED THE PADDINGTON AND FARRINGDON MAINLINE STATIONS, WAS THE FIRST RAILWAY TO OPERATE ENTIRELY UNDERGROUND. ITS LOCOMOTIVES WERE FITTED WITH SPECIAL EQUIPMENT TO DIVERT THE EXHAUST STEAM THEY GENERATED BACK INTO THEIR WATER TANKS.

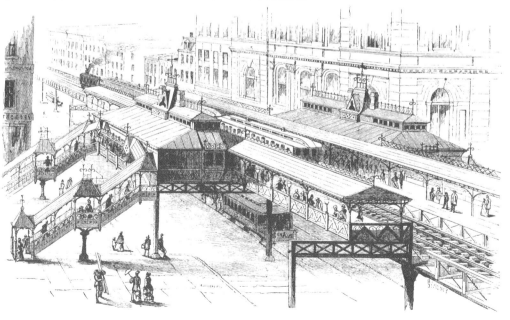

Oar and sail power
Transport over water

As the Egyptians built the pyramids they used large barges, drifting with the Nile current, to transport heavy blocks of stone for journeys of more than 800km (500 miles). Woven papyrus and reeds were their favoured boat-building materials. However, they also constructed ships from wood, binding the planks together with leather or papyrus thongs passed through holes made in their edges. A 43m (142ft) ceremonial river boat which was found buried near the Great Pyramid of Giza is the earliest known example of a ship made from planks of wood. It was built of cedar and sycamore in about 2600 BC.

Racing became a competitive sport in the early 17th century, when an annual event was set up in Amsterdam. The competitors in Britain's first race, held in 1661, were Charles II and his brother James, who raced their yachts *Katherine* and *Anne* between Greenwich and Gravesend.

Oars were used for propulsion in Egypt by 3000 BC, although sails may have been auxiliary sources of power. For steering, the Egyptians used oars at the stern. Rudders appear to have been invented in China, where they were certainly known by the 1st or 2nd century AD, although they did not appear in Europe for another 1,000 years.

In the calm waters of the Mediterranean, the oar was the ruler of the sea for centuries. However, it could not provide sufficient power for exploring other continents. That was achieved in the slow but seaworthy caravels, 15th and 16th-century sailing ships which weighed around 100 tonnes.

WORLD TRAVEL

The caravel had square sails, three masts and a simple deck below which a mid deck ran like a shelf around the inside of the hull. When he embarked on his journey westward across the Atlantic in 1492, the Genoese explorer Christopher Columbus had two caravels. But his flagship *Santa Maria* was a *nao*, a cargo vessel which measured a mere 24m (80ft). Some 27 years later, the Portuguese explorer Ferdinand Magellan's *Vittoria*, the first ship to sail around the world, was no larger.

The sailing ship reached its peak between the 17th and 19th centuries in the form of the bluff East Indiaman, developed by the Portuguese and Dutch and used for trade between southern Asia and Europe, and the elegant sailing clipper, which was built for speed and could be driven hard. At night clipper sailors did not reduce sail and 'snug down', but pressed on through the dark in the race to be the first ship home bearing the new season's crop of tea from China. From port to port they could make an average speed of about 6 knots.

Motorised boats
Powering through the waves

Steamboats driven by paddle wheels were the first motorised craft. They included an experimental model built in France in 1783 and an American ship used to ferry passengers along the Delaware river in 1790. For long sea crossings however, bigger ships, which could carry sufficient coal, were needed.

The *Great Western*, an ocean liner built by the British engineer Isambard Kingdom Brunel, was the first steamship to cross the Atlantic regularly, and it took just over 15 days to complete its maiden voyage in 1838. It was 72m (236ft) long and the 23m (75ft) saloon was far grander than that of any previous vessel. The 240 passengers were divided almost equally between first class and steerage.

BIG SHIPS

In 1835 a Kent farmer, Francis Pettit Smith, invented the screw propeller, first used by the 237 tonne *Archimede*, built in 1839. Brunel's iron-hulled *Great Britain*, launched in 1845, was the first screw-propelled transatlantic passenger ship. At some 3,270 tonnes and 98m (322ft) in length, the ship was huge by the standards of the day but was later dwarfed by Brunel's 1858 liner the *Great Eastern*, which was nearly 213m (700ft) long and weighed 18,914 tonnes.

The steam turbine, in which highly pressurised steam is directed against or through vanes on a rotor, was designed by the engineer Charles Parsons. It made its debut during the 1897 Spithead naval review, fitted onto Parsons's tiny launch, aptly named *Turbinia*. Capable of 34½ knots, the *Turbinia* caused astonishment when it outpaced some of the Royal Navy's finest steamships.

In 1901 the first big ship, the Clyde passenger steamer *King Edward*, was fitted with a turbine. Three years later the first ocean liner, the *Victoria*, was built, closely followed by the Cunard ships *Carmania*, in 1905, and the *Lusitania* and *Mauretania*, both in 1907.

The principle of using underwater 'wings' to lift the hull of a boat clear of the surface as the craft moves through the water, reducing drag and improving speed and fuel efficiency, was established by the British engineer Thomas Moy in 1861. The first commercial hydrofoil, built in the late 1930s, was used for trips along the Rhine. The hovercraft was held above the water on a cushion of air, a principle patented by the British engineer Christopher Cockerell in 1955. It was first used for cross-Channel journeys in 1968.

SHIPPING NEWS

❋ In 1624 James I watched the first demonstration of a submarine. Built in London by the Dutch physicist Cornelius Drebbel, the craft had a wooden framework covered with greased leather. It was powered by 12 oarsmen whose oars extended through sealed ports.

❋ Of the many barges used in Britain, the best known is the narrow boat, developed in the 1760s by the land agent John Gilbert and the canal builder James Brindley. This versatile craft was made specially to fit though Brindley's canal lock and measured 22m x 2.1m (72ft x 6ft 10in).

❋ The first boat to be powered by a diesel engine was a French canal boat, the *Petit Pierre*, in 1902.

First flight

' ... the motion was wonderfully smooth – smoother yet – and then! Suddenly there had come into it a new, indescribable quality – a lift – a lightness – a life!'

GERTRUDE BACON, THE FIRST BRITISH WOMAN TO FLY AS A PASSENGER, AUGUST 1909

Men first took to the skies lashed to kites. The date was around 1000 BC and the place China, but whoever thought of attaching a person to this primitive 'flying machine' remains a mystery. However, records show that condemned men were used as test pilots for man-bearing kites by the Chinese emperor Wen Hsuan Ti in AD 559. In one of his vivid descriptions of Chinese life in the 13th century, the explorer Marco Polo wrote: 'They find someone stupid or drunk for no wise man, or undepraved would expose himself to that danger.' Icarus, of Greek legend, personified classical fantasies of flying. He soared into the sky, but flew too close to the Sun and plummeted into the sea after the wax used to make his wings melted. Early attempts at gliding turned dreams to reality, but injuries and fatalities were frequent. Eilmer, the 11th-century 'Flying Monk' who broke his legs after fitting himself with wings and jumping from Malmesbury Abbey, was luckier than most.

Up, up and away

A breakthrough came in 1783 when two paper-makers, the brothers Etienne and Joseph Montgolfier, demonstrated the first hot-air balloon, made of cloth and lined with paper. A burner, fuelled by straw and wool, heated the air, which rose as it warmed, filling the balloon and lifting it upwards. It flew about 2.5km (1½ miles) before crashing and bursting into flames.

A few months later, after a second experiment in which a sheep, a rooster and a duck survived a short flight and a bumpy landing, a young physician, François Pilâtre de Rozier, and a nobleman, the Marquis d'Arlandes, became the first men to make a balloon ascent. In 1783 they took off from the grounds of the Château de la Muette in the Bois de Boulogne west of Paris. They remained aloft for 25 minutes and landed safely, having travelled about 8km (5 miles). Finally human beings could fly.

Much was achieved in balloons, including the first aerial crossing of the English Channel in January 1785 by the French balloonist Jean-Pierre Blanchard and the American Dr John Jeffries, and a parachute descent in 1797 by the French aeronaut André Jacques Garnerin. Balloons were also used for reconnaissance flights by the French Republican Army in 1794.

Powered airships were pioneered by Frenchman Henri Giffard in 1852, but the future lay in the technology of heavier-than-air machines. From basic principles described by the British engineer Sir George Cayley in an 1809 paper *On Aerial Navigation*, the German engineer Otto Lilienthal produced a series of hang-gliders of increasing sophistication in the 1890s.

Inspired by the glider pioneers and the recent introduction of the internal-combustion engine, two American bicycle manufacturers, brothers Wilbur and Orville Wright, experimented with powered flight. With each design breakthrough, the Wrights checked their advances against their knowledge of bird flight. They made four flights in 1903 on their first aircraft, the biplane *Flyer*, at Kitty Hawk in North Carolina. The longest flight lasted just under a minute and covered a mere 259.6m (852ft).

Flying machines

The following year the brothers built an improved version, *Flyer II*, and proved they were truly in control of the plane by completing a circular route. Yet the sole report of this feat appeared in a beekeepers' magazine, *Gleanings in Bee Culture*, whose editor Amos Root had been present.

By the outbreak of the First World War, the flying machine was firmly established. The hostilities led to the construction of purpose-built fighters and bombers, and in 1919 the English pilots Captain John Alcock and Lieutenant Arthur Whitten Brown used a Vickers Vimy bomber for the first nonstop flight across the Atlantic. The journey, from Newfoundland to Ireland, took 16½ hours.

Propeller-driven aircraft reached their limits of speed in the Second World War. The gas turbine, a type of internal-combustion engine which was developed independently by Frank Whittle in Britain and Hans von Ohain in Germany, offered a new source of power. Although Whittle was the first to patent the gas turbine, in 1930, it was von Ohain who was given the necessary support to develop the jet, from the German aircraft designer Ernst Heinkel. By the end of the war both Britain and Germany had jet fighters in service; the Royal Air Force's Gloster Meteor 1 and the Luftwaffe's Messerschmitt 262 first flew in 1944.

Passenger aircraft
The sky's the limit

During the First World War the combatants produced almost 200,000 aircraft. After the 1918 armistice, planes could be bought cheaply and military pilots found employment providing air displays or offering demonstration rides to thrill seekers: in June and July 1919 alone, some 10,000 'joy-riders' flew over Blackpool. Others decided that the future lay with commercial travel.

Germany's Deutsche Luft-Reederei established the world's first passenger airline in February 1919. The following month the French started a route from Paris to Brussels. In the same year the first daily international service was set up by a British company, Air Transport and Travel, linking London and Paris from Mondays to Saturdays. On its inaugural flight the plane carried consignments of newspapers and leather, several brace of grouse, some cream and a single passenger, George Stevenson-Reece of the London *Evening Standard*.

BETWEEN CONTINENTS

In the 1930s the seaplane enjoyed a brief heyday for intercontinental flights. Frequent refuelling stops and the lack of airfields in Latin America, Europe and Asia gave the advantage to an aircraft that could land on a lake or harbour. In 1937 Pan American used the Sikorsky S-42 Clipper for transatlantic trial flights, launching a regular service in 1939 with the 70-seater Boeing 314 Clippers. Its New York–Lisbon flight took 29 hours. From 1938 Imperial Airways flew the Short S-23 C class Empires from Southampton to the Far East.

THE PEOPLE'S PLANES

By the end of 1919, little more than a decade after powered flight had become a reality, Britain had five infant airlines, Germany had seven and France eight. Passenger services did not begin in the USA until 1926, but by 1929 more people were travelling by plane in that country than any other.

Converted bombers were used as passenger aircraft at first, but purpose-built airliners were soon made, starting with the British de Havilland DH16 in 1919. The 1926 American Ford Trimotor and 1930 German Junkers 52 were made entirely of metal, a novelty at the time. Most notable of the interwar designs was the American Douglas DC-3, put into service in 1936. The aeroplane seated 21 passengers, had a maximum speed of 320km/h (200mph) and an immensely tough structure.

The first pure jet airliner was the de Havilland Comet, introduced by the British Overseas Airways Corporation (BOAC) in 1952. It provided smooth flights at 805km/h (500mph) for 36 passengers but it was withdrawn after metal fatigue led to the loss of two aircraft over the Mediterranean.

The Boeing 707 set the standard for long-haul airliners after 1958. It had 179 seats and a cruising speed of about 910km/h (565mph). 'Jumbo jets', such as the 1970 Boeing 747, had even greater capacity, seating about 500, and ushered in an era of mass plane travel.

CHAPTER 7

LEISURE AND SPORT

When Parisian audiences attended the first public screening of a film in 1895 some were so frightened by the sight of a train rushing towards them that they took cover under their seats. Fortunately for film makers, such fears were shortlived and a trip to the cinema fast became one of the most popular leisure pursuits of the 20th century.

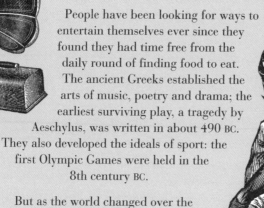

People have been looking for ways to entertain themselves ever since they found they had time free from the daily round of finding food to eat. The ancient Greeks established the arts of music, poetry and drama; the earliest surviving play, a tragedy by Aeschylus, was written in about 490 BC. They also developed the ideals of sport: the first Olympic Games were held in the 8th century BC.

But as the world changed over the centuries, so did leisure pursuits. An increase in mechanisation in the 19th and 20th centuries led to the invention of photography, sound recording, motion pictures, radio and television and a new world of mass entertainment was born.

Songs and singing
The beauty of the voice

The human voice was central to all forms of ancient music. The Book of Psalms, which instructs the faithful to 'sing unto the Lord a new song', contains some of the earliest references to Jewish singing. Chanting is also mentioned in the Rig-Veda, the first of a collection of Hindu sacred verses known as the Vedas, written in the 4th or 5th century AD but which may go back to the 2nd millennium BC. In ancient Greece, music was believed to have a significant influence on character, and instrumental sounds were considered incomplete without vocals.

Roman nurses used the soothing sounds of *lalla, lalla, lalla* to hush their charges. The oldest English lullaby to have survived – 'Lollai, lollai, litil child, Whi wepistou so sore?' – was written by an Anglo-Irish friar around 1315.

The early Christian worshippers sang their prayers in unison, with their simple melodies rooted in Greek and Hebrew musical tradition. Plainsong was the name given to this ritual melody, used by clergy and choirs to recite psalms and prayers. A later form of plainsong was known as Gregorian chant after Pope Gregory I, who ordered a review of Church music in the late 6th century.

Most of the nursery rhymes known today date from the 1600s or later. The majority were not composed for children, but are the remains of ballads, folk songs, poems ('Wee Willie Winkie'), ancient ritual ('Ladybird, Ladybird') or prayers ('Matthew, Mark, Luke and John'). The expression 'nursery rhyme' was not used in print until 1824, in *Blackwood's Edinburgh Magazine*.

Wind instruments
Flutes, trumpets and horns

The straight, pipe-like flute is the oldest known musical instrument. The earliest surviving example, found in Slovenia, was made from a bear's thighbone with fingerholes drilled into it and is about 50,000 years old.

Sculptures, paintings and pottery created from about 3000 BC show a range of wind instruments, including an oboe-like reed pipe depicted on a Mesopotamian vessel that was carved around 2600 BC. These instruments were popular in the Near East and were later adapted by the Greeks into the double-reeded aulos, the ancestor of the oboe and clarinet. The modern clarinet was developed in about AD 1690 by Johann Christophe Denner, a German instrument-maker who was trying to improve upon the *chalumeau*, a simple French wind instrument.

The transverse flute, in which air is blown sideways across the mouthpiece, was first known of in China in the 9th century BC. It reached Germany in the 12th century. Originally made of wood, it had one thumbhole and four to eight fingerholes. Keys to cover the airholes were added from 1677. In the 1800s the flute changed from a conical to a cylindrical shape.

Most trumpets in antiquity were straight, but throughout the ancient world curved animal horns were used to send men into battle and to spur them on while fighting. Long trumpets with a cylindrical tube and flaring bell were sounded by Egyptian priests and soldiers by 1400 BC. Two trumpets, one bronze and gold, one silver, were found in the tomb of Tutankhamun, the pharaoh who died in about 1340 BC.

The Romans called straight trumpets *tubas*, but also had a J-shaped model, the *lituus*, and the G-shaped *cornu*, both thought to have originated with the Etruscans, who lived in central Italy in the 7th and 6th centuries BC. Around AD 1400 instrument-makers learned to bend the trumpet's tubing into an S shape, later folding it back on itself to form a loop.

Until the end of the 18th century European horns and trumpets produced a restricted range of notes. Valves that controlled air flow more precisely were invented in about 1815. The first valved horn was patented in 1818 by the German horn player Heinrich Stoelzel. New brass instruments followed in quick succession, including the bass tuba, which was invented in 1835 in Berlin. In 1846 the saxophone was patented by the Belgian instrument-maker Adolphe Sax, who had developed it some six years earlier.

POPULAR MUSIC

Folk songs, some of which originally had a religious significance, flourished in the Middle Ages, passed down orally from generation to generation. Both folk songs and ballads were meant to accompany dancing. The ballad 'Greensleeves' was first referred to in 1580, when a licence to print 'a newe northern Dittye of ye Ladye Greene Sleves' was granted to Richard Jones.

The string section
Harps, guitars and violins

The idea of stretching a string over a hollow box so that it could be strummed and plucked probably originated in Mesopotamia. Simple bow harps can be seen in tablets from the Sumerian city of Ur around 2800 BC. Triangular frame harps, a European invention, were first shown on Celtic manuscripts of the 9th century. In about 1720 a German musician named Jakob Hochbrucker created a harp with pedals that raised the pitch of the strings by a semitone. Advances made around 1810 by Sébastien Erard, a Parisian instrument-maker, led to the greater range of tone found in the modern classical harp.

Lutes, which have strings running over a sound box and along the arm, were depicted on Babylonian seals in about 2300 BC. 'Lute' comes from the instrument's Arabic name *al-ud*, 'the wood'. A family of guitar-shaped lutes known as gitterns were the forerunners of the guitar, which was being played in western Europe by the late 1300s.

Made of horsehair strung on wood, the bow appeared in the Byzantine Empire around the 10th century. It was used to play the pear-shaped vielle, ancestor of the two main kinds of viol that developed in the 1500s: the viola da gamba, held between the legs, and the viola da braccio, played on the arm.

The violin, which evolved from the medieval fiddle, developed in Italy in the late 1500s as an instrument to provide music for dancing. The first great violin-maker was Andrea Amati. One of the pupils at his workshop was Antonio Stradivari, who before his death in 1737 perfected the art of violin-making, creating more than 1,000 instruments.

Organs and pianos
Keyboard instruments

Working in his father's barbershop in Alexandria in the 3rd century BC, the future engineer Ctesibius accidentally dropped a lead weight down a pipe and noticed that the compressed air was pushed out with a squeak. He adapted the idea into the hydraulis, a small organ built along the same lines as a set of panpipes, incorporating a series of tubes of different lengths. Air was pumped into the internal chamber, where water stabilised the air pressure; sliders controlled the release of air through individual pipes.

Perfected in the late 1st century BC by the architect Vitruvius, the organ came to be played throughout the Roman Empire.

Today's musical instruments have evolved over thousands of years from simple shells, bones or gourds. Ancient peoples modified these, for example by making holes, to create instruments that produced just one note each.

The craft of organ-building almost disappeared from Europe after the fall of Rome, but was preserved in the Middle East. Instruments occasionally arrived in the West as gifts, such as the organ that was offered to the Frankish king Pepin by the Byzantine emperor Constantine Copronymous in AD 757. By the 10th century organs had been adopted to provide church music, but it was only in the following century that the unwieldy sliders were replaced by levers and then by keyboards.

Rudimentary stringed instruments that were precursors of the piano were played before Biblical times. The psaltery, played by plucking, had strings stretched over a sound box of four unequal sides. This gave birth to the Middle Eastern dulcimer, in which the strings are struck with hammers.

The first person to combine the action of the dulcimer with a keyboard was Bartolommeo Cristofori, an Italian instrument-maker. His invention is listed – under the name *arpicembalo* – in an inventory from 1700 of the Medici family's instruments; a few years later it is referred to as a *gravicembalo col piano e forte*, or 'harpsichord with soft and loud'. The earliest known composer to have written pieces for the piano was the Italian Lodovico Giustini, who published 12 piano sonatas in 1732.

Percussion instruments
Drumming out a rhythm

The first drumbeats may have been sounded out on mammoth skulls and animal skins stretched over bones in the Ukraine around 14,000 BC. The oldest drums to survive intact are clay examples from about 3000 BC found in Germany and the Czech Republic. Vases and pots of the same period reveal that drums were played in the Near East; crocodile-skin drums of a similar age have been unearthed in graves in China.

Kettledrums – single-headed drums shaped like a bowl – were introduced into Europe from the Near East by the Crusaders in the early 1100s. These 'nakers' as they were known, from their Arabic name *naqara*, survive as modern timpani. In the 19th century the addition of a handle enabled rapid tuning, and allowed pitch to be altered during play, making the timpani central to the orchestral percussion section.

The orchestra
Musicians in concert

One of the earliest depictions of a group of musicians is on a Sumerian seal made around 4000 BC, which shows a small orchestra based on kettledrum, harp and horns. Musicians flourished in the Near East. The orchestras that entertained the pharaohs during the 2nd millennium BC were often made up of Near Eastern musicians, many of them women. Most musicians earned a living by playing at banquets. Harps, lyres and lutes were played alongside oboes, flutes, tambourines and rattles.

The Western orchestra is rooted in the Renaissance. Because orchestras were needed to accompany the singers, the development of opera in the 1600s stimulated polyphonal instrumental music, combining two or more melodic parts. By 1800 the baroque ensemble or orchestra was established, founded on string instruments, recorders or flutes, oboes, bassoons, keyboard instruments, trumpets, horns and cello or bassoon.

Musical notation
A lyrical alphabet

When cuneiform tablets found in the Babylonian city of Ur in the 1920s were analysed, it became clear that one, from the late 1st millennium BC, held the earliest known representation of a musical scale. Another tablet of 1400 BC, found in Syria, contained the notes and text of a song. Ancient music had been thought to consist of one melody, but the tablet showed several harmonies.

Ancient Greek musical theory was the starting point for modern musical notation. The existence of fixed ratios between intervals of sound was discovered by the mathematician Pythagoras in the 6th century BC. This led to a systematic means of depicting notes and by 450 BC scores were being written for plays such as Euripedes' *Orestes*.

The modern system of notation evolved from church music in the 9th century, when plainsong was written using dots and squiggles known as neumes, from the Greek *neuma*, 'gesture'. Neumes were not a scale system, but appeared above a text to indicate changes in pitch.

A six-note scale and the stave, a set of horizontal lines on which the note symbols were placed, was invented in the 11th century by an Italian monk, Guido of Arezzo. By the late 1600s his scale had been replaced by a seven-note one.

In around 1260 the musician Franco of Cologne created a series of shapes to denote the duration of sounds. Scores became more precise with the introduction of breves and semibreves in the 13th century, minims and crochets in the 14th, and quavers and semiquavers in the 15th.

KEEPING TIME

✷ The first reliable metronome was devised in Amsterdam in 1814 by the German organ-maker Dietrich Winkel, but produced by the musician Johann Nepomuk Maelzel, who patented it in his own name. Winkel sued successfully, but Maelzel had retired to the USA with the profits.

✷ Karaoke machines are generally believed to have originated in Japan in 1976, but Roy Brooke of Cheshire claims to have sold a similar machine to some Manchester clubs in 1975.

The rituals of dance
Rhythms of life

To ancient peoples dancing was an important means of expression, first depicted in cave paintings in Africa and southern Europe some 20,000 years ago. Descriptions dating to 1800 BC relate that the Egyptians danced to Osiris, the god of vegetation, to ensure the fertility of the Nile plains. Wealthier families entertained their guests with dancing slave girls. To the ancient Greeks, dance was a pastime of the gods and a vital part of education and military training.

Folk dancing is a purely recreational pursuit rather than a ritual performance. Nevertheless it has roots in religious traditions. Dancing around a sacred object was an ancient form of religious observance later incorporated into folk dances. In Europe, for example, dancing around a maypole recalls the ancient rituals of tree worship associated with fertility.

As religious calendars were set, celebrations often coincided with pagan festivals, as with Christian Easter and the rites of spring. Morris dancing, associated with Whitsun, probably evolved from ritual sword dances, widespread throughout Europe, which had their origins in nature worship.

In Europe folk dancing was largely an entertainment for peasants until the 14th century, when people from all classes began to join in. English country dances originate from various May Day ceremonies. In the early 1700s social dances began to develop national identities, as with Scottish jigs, reels and strathspeys. From English dance steps and the French quadrille, pioneers who emigrated to the New World created American square dancing.

Ballet dancing
The terpsichorean arts

Ballet had its origins in the 14th century when formal versions of folk dance were developed in the royal courts of Europe. For anyone attending the courts of the Italian and French Renaissance, dancing skills were compulsory and grand families had their own dancing masters. These teachers also began to publish books – the earliest was *De arte saltandi et choreas ducendi* (*On the Art of Dancing and Conducting Dances*) by Domenico of Piacenza, published around 1420.

Many formal dances were performed as interludes in operas or plays. Known as *ballets de cours*, French for 'court ballets', their steps and sequences descended from folk dances and court processions and were performed by aristocratic amateurs.

The first true ballet de cour was *Circé ou le ballet comique de la reine* (*Circe or the Comic Ballet of the Queen*), which was shown in Paris in 1581. When the French king Louis XIV founded the Académie Royale de Danse in Paris in 1661 in order to 're-establish the art in its perfection', he emphasised the divide between refined court and peasant dance. But it was not until 1672 that Louis recognised the need for purely professional dancers.

Initially male dancers predominated, and it was thought unseemly for ladies of the court to dance with them. The first professional ballerina, Mademoiselle de Lafontaine, appeared in *Le Triomphe de l'amour* (*The Triumph of Love*) in Paris in 1681. With music by the Italian-born French composer Jean Baptiste Lully, it was the first ballet to tell a tale through dance and gesture.

By 1713 the first fully professional dance school had been established at the Paris Opéra, and steps such as the entrechat and chassé formalised and named. The first ballet dancers to perform the pas de deux, or 'step for two', were also the first to dance on pointe (on the tips of their toes) in a staging of Charles Didelot's *Flore et Zéphyre* in London in 1796. They were attached to lifting devices, but such equipment was soon dispensed with as unnecessary.

TAP DANCING

Homesick African slaves, shuffling barefoot in time to work songs and blues, may have put metal bottle tops in between their toes as a way of adding sound to their rhythms. Tap dancing, an American form of theatrical dance that added elements of ritual African foot stamping, Irish jigs and European clog dances, was performed by black Americans in their 19th-century minstrel shows. From 1925 metal taps were attached to dancers' shoes, but tap first took the world by storm in the 1930s and 1940s, when it was transformed for the big screen by such maestros as Fred Astaire and Gene Kelly.

From ballroom to disco
Dancing together

Popular dance was dominated by circle and line formations until the *danse à deux*, 'couple dancing', became standard in the 1300s. Thereafter, countless varieties of dance emerged. The waltz, named from the German word *walzen*, meaning 'to rotate', appeared in the 18th century. It developed from the Ländler, a traditional dance for couples from southern Germany and Austria with the familiar 1-2-3, 1-2-3 rhythm. When first performed by twirling embracing couples, the waltz sent shock waves through polite society.

More shocking yet was the tango, born in about 1880 in the slums of Buenos Aires that combined a Spanish flamenco-type dance with the raunchy Argentine *milonga*. By the outbreak of the First World War the tango had become a craze throughout Europe. The foxtrot also originated around this time, probably named after the American comedian Harry Fox in 1913. His act in the Ziegfeld Follies included a series of trotting steps.

The dominance of couple dancing was broken by jazz dance, based on rhythms taken to the USA by West African slaves. At the turn of the 20th century such dances as the black bottom appeared in the black American dance halls of Atlanta, in Georgia. During the 1930s these evolved into the acrobatic jive and jitterbug, in which dancers improvised freely on a few set steps.

By 1962 'disco', from the French *discothèque*, or 'record library', referred to dimly lit Parisian dance halls where people danced the twist. Devised in the USA in about 1958, the twist was initially banned in dance halls, but went on to gain widespread popularity.

Discos had arrived in Britain by the mid 1960s, playing mostly Tamla Motown or soul music. Music created with discos in mind, and the dancing style that accompanied it, originated in the 1970s, encouraged by films such as *Saturday Night Fever*.

On the stage

'If a play does anything – either tragically or comically,
satirically or farcically – to explain to me why
I am alive, it is a good play.'

TYNAN *LEFT AND RIGHT*, KENNETH TYNAN, THEATRE CRITIC, 1967

Religious ritual inspired the first dramas. In the 4th millennium BC Egyptian players dressed in lion face masks and other costumes to pay homage to the god Bes, protector of pregnant women. The first dramatists, in a modern sense, were the ancient Greeks, whose plays addressed such human emotions and failings as passion, the lust for power, and self-destruction. Greek tragedy evolved from the choral lyric, a poem, often religious, sung and danced by a chorus. Dialogue and dramatic action are said to have been introduced by Thespis, probably an actor and playwright. He added a single actor who stood out from the chorus, playing several parts and addressing the chorus directly. Direct exchanges between characters began when the playwright Aeschylus added a second actor. His tragedy *The Suppliant Women*, written in about 490 BC, is the earliest play to have survived complete.

Thespis was also the first known recipient of a theatrical award, presented at the Athenian festival of Dionysus in 534 BC, as well as the source of the word 'thespian'. 'Tragedy' comes from the Greek *tragos*, 'goat', a reference to animal sacrifices made at festivals or to the practice of giving a goat as a prize, and *oide*, meaning 'song'.

Historical drama

Comedies first entered theatrical competitions in 486 BC. Derived from the Greek *komos*, 'revel', and *oide*, comedy developed from the revels associated with fertility rites. By the 3rd century BC comedies were mostly political satires and burlesques featuring some 30 stock characters such as the lover, the bounder and the virtuous woman.

The great theatres built by the Greeks were inspired by the arenas of Bronze Age Crete, where the Minoans enjoyed spectator sports in amphitheatres with stepped seating. In the 6th century BC the Greek theatre of Thorikos consisted of wooden benches set in the hillside looking down on a rectangular *orkhestra* where the chorus danced.

By the mid 5th century BC the theatre of Dionysus near the Acropolis in Athens included a *skene*, a wooden façade with three doors. This provided the background for the actors – and the derivation of the word 'scene'. The Greek playwright Sophocles introduced the idea of showing the location of the action by fitting painted canvasses onto the *skene*. The Romans dispensed with the *orchestra* and the chorus, presenting their comedies on a high stage across which a curtain could be drawn.

The early Church condemned entertainment of this kind, closing all theatres in the 6th century AD before later realising the potential of drama to encourage moral behaviour. In western Europe the earliest recorded drama dates to the mid 10th century, when plays based on the life of Jesus formed part of the Church's system of public worship. From the 13th century medieval mystery and miracle plays related stories from the Bible and the lives of the saints, while morality plays personified vices and virtues.

Secular theatre survived in Europe in the itinerant groups of entertainers who performed at markets and fairs, as well as in the minstrels employed by lords. From the late 15th century in England, small companies of actors were hired to give 'interludes' – short dramatic sketches – between the courses of banquets held in the great houses of the nobility.

Paid performers

Professional actors and, later, actresses emerged in Europe from the mid 16th century. In Italy they were first seen in the commedia dell'arte, semi-improvised satires and comedies in which masked actors played a standard cast of characters, including flirtatious Columbina, miserly Pantalone and quick-witted Arlecchino (Harlequin).

After Henry VIII broke with the Catholic Church in 1532, English writers satirised the old religious plays and explored social and political issues. By the end of the century professional playwrights such as Christopher Marlowe, Ben Jonson and William Shakespeare were writing for companies who performed in purpose-built theatres rather than inn yards.

England's first playhouse was erected in 1576 by James Burbage, leader of a company of actors under the patronage of the Earl of Leicester. Burbage put up a timber-framed structure with a thatched roof and three galleries in Shoreditch, east of London's city walls, and named it the Theatre.

Although actresses had appeared in Britain in the 1570s with visiting Italian commedia dell'arte players, women first began to appear on the public stage only after the Restoration of Charles II in 1660. Previously they were seen only in court masques.

Magic shows
Tricks of the trade

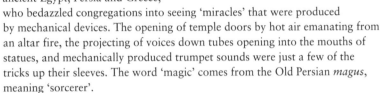

The first of the great illusionists were the sorcerer-priests of ancient Egypt, Persia and Greece, who bedazzled congregations into seeing 'miracles' that were produced by mechanical devices. The opening of temple doors by hot air emanating from an altar fire, the projecting of voices down tubes opening into the mouths of statues, and mechanically produced trumpet sounds were just a few of the tricks up their sleeves. The word 'magic' comes from the Old Persian *magus*, meaning 'sorcerer'.

An ancient Egyptian papyrus relates how the pharaoh Cheops, who ruled around 2600 BC, was entertained by a magician who could 'put on again a head that hath been cut off', proving it with a goose. Sword swallowing, neither an illusion nor a trick, was known in ancient Greece and Rome. A Roman physician noted in the 2nd century AD that *abracadabra* worked as a charm to cure fevers, but it may have originated as the Greek *abrasadabra*, a magic word used by some early Christian sects.

In the 1700s the terms 'conjurer' and 'magician' lost their supernatural associations and performers of magical illusions began to play in theatres and in music halls. Mind-reading was introduced as a trick in 1781 by Philip Breslaw. The illusion of sawing a woman in half was first performed in London by the British magician P T Selbit in 1921.

Puppet performances
Punch & Judy

The crotchety nature of Punch may have evolved from the ungainly, deceitful country bumpkin who was a stock character of Greek and Roman mimes. Puppet shows were also enjoyed in Greece in the 6th century BC.

When, in the 1550s, the commedia dell'arte, 'comedy of art', developed in Italy, one of its stock characters was Pulcinella, a sly, braggart servant who always managed to escape punishment. In the early 1600s Pulcinella became a favourite of Italian marionette shows.

In 1662 the diarist Samuel Pepys noted Pulcinella's arrival in England. Renamed Punch, he took characteristics from the English clown while gaining a nagging wife called Joan and, probably, his humped back. English morality plays were also the source of Punch's fight with the Devil – replaced by a crocodile in the mid 19th century.

But in the late 1700s his popularity declined. With all their paraphernalia, marionette shows became uneconomical, so Punch was reborn as a glove puppet – a form first seen in England in the 1400s, when puppet shows were based on mystery plays. The text of the shows was established around 1800 – although some of the characters have come and gone, the basic story has remained the same – and by 1820 Punch's looks and his habit of beating his wife, now called Judy, were set, as was the expression 'pleased as Punch'.

Opera and musicals
Tuneful tales

The idea of telling a story in words set to music originated in Italy in the late 1500s. The Camerata, a group of musicians, poets and aristocrats in Florence, helped to create a new form of musical drama called *opera in musica*, or 'work in music'. The first true opera was written by two members of the Camerata, librettist Ottavio Rinuccini and composer Jacopo Peri: *Dafne* was performed as part of the Florence Carnival in 1597.

'Through-sung' opera, in which all or most of the libretto is set to music, reached London in 1656 with one of the few theatrical performances of the Commonwealth period. Sir William D'Avenant's *The Siege of Rhodes* was set to music by a committee of composers. It was the first time movable perspective scenery was used in a British theatre, and also the first dramatic performance to feature a woman, Mrs Edward Coleman. The British continued to prefer 'semi-opera' – an amalgam of spoken dialogue and singing – until the early 1700s when through-sung opera caught on. In 1710 the Queen's Theatre, Haymarket, became London's first opera house and German-born George Frideric Handel staged *Rinaldo* there the following year.

POPULAR PANTOS

The pantomime, named from the Latin pantomimus, meaning 'complete mime', was established in Roman times as a light entertainment in which a chorus narrated the plot while actors mimed the action. Its modern British form was adapted from harlequinades, comic antics popular in 17th-century France, by the British harlequin John Rich in the early 1700s. The tradition of basing story lines on fairy tales began in the late 18th century, but the Victorians introduced the convention of a woman playing the principal boy.

MUSICAL MILESTONES
When church organs were banned by the Puritans in the mid 1600s many of them were bought by publicans, who then employed organists to entertain their customers. By the early 1800s taverns and 'song-and-supper rooms' provided food, drink and a succession of comedy turns, singers and acrobats. Such entertainment proved so popular that in 1852 an English publican named Charles Morton built the first of London's great music halls, the Canterbury. In the late 1800s the music halls were rebuilt as theatres, without supper tables and became known as 'palaces of variety'.

The American equivalent of music hall came to be known as vaudeville, a term introduced by French immigrants possibly from the words *vau* or *val de Vire*, referring to the valley of Vire in Normandy, known for its popular songs. The first vaudeville show, which offered eight acts, including comics, singers and dancers, was devised in 1881.

From the late 1800s music hall and vaudeville faced competition from musical comedies and revues. William Gilbert and Arthur Sullivan had been delighting audiences with their light operas since 1871. But in 1892 Osmond Carr's *In Town*, the first English musical comedy, was staged in London. As plots grew stronger and were integrated more with the music, the word 'comedy' was dropped. The American musical gained its first major London success in 1898 with Gustave Kerker's *The Belle of New York*.

Capturing a likeness

'From today painting is dead.'

FRENCH PAINTER PAUL DELAROCHE, 1839, ON HIS FIRST SIGHT OF PHOTOGRAPHS

The story of the camera began with Arab astronomers of the 9th century: they knew that a beam of light reflected from an illuminated object and entering a darkened room through a small hole would project an image of the object upside-down on the opposite wall. They used this principle to observe sunspots and eclipses without damaging their eyes. The *camera obscura*, named from the Latin for 'dark chamber', was little known in Europe until the 15th century, when it was described by the Italian artist and scientist Leonardo da Vinci. In the 17th century the room became a movable tent, then a portable box.

However, how could such images be recorded and preserved? The key to progress lay in the way that silver nitrate darkens on exposure to light, a phenomenon discovered in 1727 by the German physicist Johann Heinrich Schulze. Thomas Wedgwood, son of the pottery-maker, and the chemist Humphry Davy were probably the first to make use of this phenomenon later in the century. But they could not 'fix' the images, which turned black when exposed to more light.

First photos

In the 1820s the camera, light-sensitive chemicals and a method of stabilising the image were brought together by Joseph Nicéphore Niépce, a French chemist. He captured the oldest surviving photograph from his house in Burgundy. The indistinct view is captured on a pewter plate coated with bitumen of Judaea (a type of asphalt), which hardened white in the light. From the positions of the shadows, it is clear that the exposure time was about 8 hours.

Niépce went into partnership with a theatrical scene-painter Louis Jacques Mandé Daguerre. It was Daguerre who, in 1837, succeeded in using common salt to fix a silver image on a polished and silvered copper plate. Each of the solid metal plates used by Niépce and Daguerre was a 'one-off', with a mirror-reversed image.

Two years earlier Daguerre had realised that an image was forming on the plate before it was visible to the eye. His discovery is said to have happened after he left an underexposed plate in a cupboard, planning to repolish then reuse it. To his surprise, when he took the plate out an image was visible. He repeated the 'accident' until he discovered that a chemical had brought out, or developed, the hidden or 'latent image': it was apparently mercury vapour, which had seeped into the cupboard from a broken thermometer.

Daguerre now realised that exposures could be reduced to 2 or 3 minutes, a reasonable time to expect humans to keep still. Probably the first person to be photographed was a man having his boots blacked on a Paris boulevard, at whom Daguerre pointed his camera in 1839. Taking pictures of people became so important that more than 90 per cent of all daguerreotypes were portraits.

In the autumn of 1839 the new medium crossed the Atlantic. The first American daguerreotype, taken by D W Seager, was of St Paul's Church in New York. Within a year, the daguerreotype, which the American essayist Oliver Wendell Holmes dubbed the 'mirror with a memory', had swept through the USA.

At the same time as Daguerre was developing his process, English scientist William Henry Fox Talbot was also experimenting with photography. It was he who invented the negative-positive process, enabling copies to be made. The earliest surviving negative is of a diamond-latticed window in Fox Talbot's family home, Lacock Abbey in Wiltshire. Probably exposed for 1 to 2 hours, it was taken in August 1835. His process was unveiled at the Royal Society, London, at the end of January 1839, a few weeks after Daguerre's.

Written by light

Fox Talbot discovered the latent image for himself and, modifying his process, patented the resulting calotype, named from the Greek *kalos*, 'beauty', in 1841. The process was unable to rival the daguerreotype for its precision of detail, but was capable of much subtler effects with more gradation of tones. A friend of Fox Talbot's, the English astronomer Sir John Herschel, coined the word 'photography' in the 1840s from the Greek *photos*, meaning 'light', and *graphein*, 'to write'.

By the 1850s the calotype and daguerreotype had been superseded by new processes. Glass plates, used instead of paper to receive negative images, made both fine detail and multiplication of prints possible for the first time. However, photography remained in the hands of professionals who had the necessary knowledge of chemicals as well as the skill to perform often complex manipulations.

George Eastman, an American bank clerk who became a photographic manufacturer, revolutionised photography in 1888 with his Kodak camera, which could be used by amateurs. Once the owner had used up the 100 frames on its paper-backed roll film the whole camera was usually sent back to the Kodak factory for the pictures to be processed and a new film to be put in. However, it was Kodak's affordable Box Brownie, priced at 5 shillings (25p) in Britain, which brought photography to the masses when it was introduced in 1900.

The birth of the gramophone
Musical machines

Edouard-Léon Scott de Martinville, the Parisian painter, connected a megaphone to a stylus made of a hog bristle which rested on a revolving drum. When he made a noise, a wavy line was etched into the smoke-blackened paper covering the drum. His Phonautograph had produced the first visual trace of sound waves, in 1857.

Scott had no interest in playing back recorded sounds, but two decades later the American inventor Thomas Alva Edison, working on a device to translate Morse code signals into marks on paper, succeeded in doing so. On December 6, 1877, his mechanic, John Kruesi, built the first phonograph to Edison's design.

Turning the handle of its cylinder, Edison shouted the rhyme 'Mary had a Little Lamb' through a tube attached to a diaphragm which vibrated to the sound. This caused a stylus to make indentations in the aluminium foil that covered the cylinder. A stylus attached to a listening tube was then used to reproduce the sound.

Recordings made in 1878 included a speech by a US president, Rutherford B Hayes. But these soon wore out.

Recordings that could be played again and again were first made on a machine patented in 1886 by Alexander Graham Bell, the inventor of the telephone, his cousin Chichester Bell and the inventor Charles Sumner Tainter. The Graphophone's cylinder was covered in wax, which was engraved by the stylus. Edison immediately made his own wax-cylinder version. The rivals then joined forces, producing Edison's Perfected Phonograph in 1887.

Short and long play records
Captured on disc

Emile Berliner, a German migrant to the USA, harboured an ambition to 'etch the human voice'. Working on an idea originated ten years earlier by French scientist and poet Charles Cros, Berliner used a stylus to trace a wavy line on a glass disc covered with lampblack and linseed oil. He hardened the trace with shellac, a purified resin, and used it as a 'master' negative to transfer the pattern to a flat metal disc. The grooves on the second disc could play back the recorded sound.

Berliner demonstrated a disc-record player, the hand-cranked Gramophone, in May 1888. In 1897 the Berliner Gramophone Company produced their first shellac discs for playing on a 78 rpm (revolutions per minute) turntable.

MELODIES FOR SALE

The original jukebox, an Edison phonograph, was installed at the Palais Royal Saloon in San Francisco, USA, in 1889. Its listening tubes came into action when a nickel was placed in the slot. The name 'jukebox' comes from a 1939 article in Time magazine which recorded that the '"jukebox" … retails recorded music at 5 cents a shot in bars.' In US slang, the word 'juke' means 'to dance'.

On Berliner's early recordings titles were engraved at the centre. Printed paper labels were the brainchild of Eldridge Johnson, first issued in 1900 by his Consolidated Talking Machine Company, New Jersey. Johnson had had the idea of using a long, flared horn to amplify sound in 1898. Depicted on every label was the 'dog and trumpet' trademark, adapted from a painting called *His Master's Voice* by the English artist Francis Barraud. In the original painting, a fox terrier, Nipper, listened to the horn of a phonograph, but this was substituted with a gramophone in the trademark illustration.

Recording quality improved with the advent of electrical recording systems. Edison and Berliner had both developed microphones for converting sound into electrical energy, but the resulting sound needed amplification. Solving the problem in the 1920s, the Western Electric Company combined a microphone with public address amplifiers, and in 1925 the Brunswick Panatrope, the first all-electric record player with loudspeakers, was launched.

The US company RCA-Victor was the pioneer of long-playing records, producing a 33⅓ rpm shellac recording of Beethoven's Fifth Symphony in 1931. But after shatterproof plastic records were launched by Columbia Records (CBS) in 1948, the 78 rpm disc eventually became obsolete. Etched with 100 microgrooves per centimetre (254 per inch) for high-quality sound, and providing 23 minutes' playing time on each side, LPs were a great success. As gramophones and discs became more readily available, sales of popular music soared, and in April 1935 *The Lucky Strike Hit Parade*, featuring the week's most popular songs, began on US radio. The first 'number one' was George and Ira Gershwin's 'Soon'. After his 1941 hit 'Chattanooga Choo-Choo' sold a million, the American bandleader Glenn Miller was presented with a copy of the recording sprayed with gold – the first example of a performer receiving a gold disc.

Tape recording
Magnetic music

To record telephone conversations Valdemar Poulsen, the 'Danish Edison', devised the Telegraphone, in effect the first answering machine, by capturing sound on piano wire with magnetic recordings in 1898. By 1920 the search had begun for new recording surfaces, and in 1929 Dr Fritz Pfleumer filed for a patent in Germany for a flexible tape with a magnetic coating.

DIGITAL TECHNOLOGY

Musical watches made in Switzerland in the 18th century gave birth to the music box of the 1790s. When the lid of the box was opened, a cylinder with raised indentations turned against a small metal comb to play a tune. The original music box was digital in concept – with each indentation representing a single piece, or 'bit', of information. Modern digital recording began with the compact disc (CD), which stores sound data as microscopic pits on its surface. It was developed by the Dutch company Philips in the 1970s: the first CD players, using laser beams to scan the discs, were sold by Sony in Japan in 1982.

Pfleumer joined forces with AEG of Berlin, launching the Magnetophon, the first modern tape recorder, in 1935. German models recovered by the Allies at the end of the Second World War inspired the development of high-quality tape recorders in Europe and the USA in the late 1940s.

Wishing to listen to music when he was travelling, the founder of the Japanese company Sony, Masura Ibuka, had the idea of adding a pair of lightweight headphones to a small tape recorder. His idea lead to the development of the Sony Walkman, which went on sale in 1979.

Moving pictures

'By its means historical events can henceforth be preserved just
as they happened and brought to view again not only now,
but also for the benefit of future generations.'

OSKAR MESSTER, GERMAN INVENTOR, PRODUCER, DIRECTOR AND PROPAGANDIST, 1897

The principles behind photography, first demonstrated in 1839, led directly to the birth of motion pictures, but two other essentials of cinematography came before it. The first of these was projection in its simplest form – using light to cast shadows onto a screen or wall. Shadow puppets originated in the Far East in the 14th century and in 17th-century Europe the same principle was used in 'magic lantern' shows.

The second essential ingredient of film, persistence of vision – the retention by the brain of an image on the retina for a split second after it disappears – had been appreciated since the 17th century. It seems to have been demonstrated for the first time in 1824 by the English physician Peter Mark Roget, of *Roget's Thesaurus* fame. Soon the zoetrope and other optical toys used this phenomenon. After 1839, attempts were made to use photographs in these toys, but because of the lengthy exposure times needed, the process of taking the images proved to be painfully slow. Only after the 1870s did film-making finally become practical.

Fast forward

In the 1870s English photographer Eadweard Muybridge was commissioned by Leland Stanford, railway tycoon and former governor of California, to photograph a running racehorse, in an attempt to prove that as a horse galloped all its hoofs cleared the ground for brief periods. Muybridge's black and white photographs, taken by a series of cameras operated by trip wires set off by the horse as it ran past, showed that this was indeed the case.

Etienne Jules Marey, a French scientist, used a single camera – his 'photographic rifle' – to take pictures of birds in flight in the 1880s. But only celluloid film and projectors would enable long sequences to be taken and viewed so that motion appeared realistic. Meanwhile the best results came from mounting sequences of pictures and running them in mechanical 'flicker books'. Many were viewed in machines such as the Mutoscope, a freestanding, 'penny in the slot' peepshow invented in 1894.

In 1891 the American inventor Thomas Alva Edison patented the Kinetoscope, mainly devised by his assistant W K L Dickson. This viewing box used images captured on celluloid roll film, introduced in 1889 by George Eastman. The first Eastman Kodak film was 70mm wide. Edison split it down the middle so he could join the pieces together and make longer lengths. In 1907, 35mm film became the standard.

With roll film it was possible, at last, to record photographic sequences in continuous form. The Kinetoscope used 15m (50ft) looped rolls, each with some 20 seconds of playing time. The films were recorded at a specially built studio at the Edison Laboratories in New Jersey. Kinetoscope parlours containing coin-operated machines sprang up in the USA and Europe, but the pictures could be viewed by only one person at a time.

The world's first public screening of a film was in the basement of the Grand Café, Paris, on December 28, 1895. The event had been organised, and the films shot, by the brothers Auguste and Louis Lumière, whose family made photographic equipment. Their Cinématographe incorporated both camera and projector and was named from the Greek *kinema*, or 'motion'. It used a claw mechanism to hold each frame in front of the lens for a split second before moving the film on. Ten films were shown, each 1 minute long, including workers leaving a factory, a family mealtime and a blacksmith at work. Many viewers were so frightened by seeing a train steaming towards them that they hid under their seats.

Hooray for Hollywood

Cinema was taken up most enthusiastically in the USA. *The Great Train Robbery*, often considered the first true movie, was a narrative western lasting 11 minutes and made in 1903. By 1907 film-makers were starting to find the sunshine and scenery of southern California ideal for their purposes: Nestor film studios was the first to make a permanent base in Hollywood.

Sound and pictures were incorporated into Edison's Kinetophone, a combination of his phonograph and Kinetoscope, of about 1895. Synchronised dialogue first accompanied projected film at the 1900 Universal Exhibition in Paris, where films included scenes from such dramas as Shakespeare's *Hamlet*, and featured famous stage actors such as Sarah Bernhardt. However, tone quality was generally so bad, and the synchronisation so difficult, that most films were 'silent' until sound-on-disc and sound-on-film systems – the 'talkies' – were introduced in the 1920s.

The birth of television
Pictures on 'the box'

IN MARCH 1930 RAMSEY MACDONALD, THE BRITISH PRIME MINISTER, WAS PRESENTED WITH A TELEVISION RECEIVER FOR USE IN 10 DOWNING STREET BY THE SCOTTISH ELECTRICAL ENGINEER JOHN LOGIE BAIRD. AFTER WATCHING *THE MAN WITH THE FLOWER IN HIS MOUTH*, BY THE ITALIAN PLAYWRIGHT LUIGI PIRANDELLO, MACDONALD WROTE TO BAIRD ON APRIL 5: 'YOU HAVE PUT SOMETHING IN MY ROOM WHICH WILL NEVER LET ME FORGET HOW STRANGE IS THIS WORLD AND HOW UNKNOWN.'

The name of John Logie Baird is indelibly linked with the invention of television, but no one person was responsible. It developed from a series of discoveries and innovations, beginning with Abbé Castelli, an Italian-born priest working in France, who found, in 1862, that variations of light could be transmitted down a telegraph wire as electrical impulses.

The first real step forward came in 1884 when a German student, Paul Nipkow, had the idea for an 'electrical telescope'. It had a rotating disc pierced with holes which 'dissected' images mechanically into a series of separate lines of varying light intensity. The light was converted into an electric current then projected onto a screen through a second rotating disc.

By 1897 a second German, the physicist Karl Ferdinand Braun, had solved the same problem in a different way. He developed the cathode-ray tube, inside which streams of high-energy electrons passed through a vacuum and created spots of light when they hit a fluorescent screen at the far end of the tube. This method of scanning electrically, rather than mechanically, was the technological basis of modern television.

CAMERA MAN

No further progress was made until the development of the first television camera, which could scan an image using an electron beam. Its inventor was Vladimir Kosma Zworykin, a Russian-born American, who applied for a patent for his iconoscope in 1923, and spent ten years perfecting it.

John Logie Baird had built his first, but unsuccessful, apparatus on an old tea chest, with a disc made from a hatbox lid, mounted on a knitting needle and driven by a motor from an electric fan. The clarity of the images produced by his later working mechanical systems was poor. He sharpened the images by breaking them up into more lines. Later, in an attic above a Soho restaurant in London, with equipment which included a Nipkow's disc, Baird produced the first image of a living face. Early in 1926 he used his device, called the Televisor, to show such images to members of the Royal Institution.

Baird went on to produce a colour transmission system which he demonstrated in 1928. In the same year he made the first transatlantic television broadcast, transmitted from London to Hartsdale in New York. Baird had secured his place in television history, but his crude mechanical system was destined to be overtaken by electronic imaging based on Zworykin's iconoscope. Zworykin was later hailed as the father of modern television.

Broadcasting begins
Programmes for the public

John Logie Baird's enthusiasm, and his genius for publicity, led to the realisation of television's potential for mass entertainment. The Baird Company opened a television studio in London in 1928. Later that year the world's first scheduled television service was launched in the USA by the General Electric Company's Station WGY Schenectady, New York. Their first transmissions were of the faces of men talking, laughing and smoking.

Baird's approach made use of singers and other performers. His company dominated early programme making in Britain, operating a daily broadcast service after securing the use of the BBC's transmitter, in 1929. But it was not possible to broadcast pictures and sound simultaneously from one transmitter. Pictures of singers mouthing words alternated with a black screen over which the sound was broadcast. On the day Ramsay MacDonald acquired his receiver a second transmitter came into use; Annie Croft and Gracie Fields sang on the first simultaneous sight and sound broadcast.

In Tokyo in 1931, Japanese engineers broadcast a baseball game on closed circuit television. That June British viewers experienced their first outside broadcast – Baird's live transmission of the Derby. Most such programmes were received by enthusiasts who built their own sets using diagrams presented in *Television*, a new specialist magazine.

The video revolution
Entertainment at home

In 1927 John Logie Baird became the first person to attempt to create a video. Images were recorded on 78 rpm discs which could then be played back on a gramophone linked to his Televisor. However, Baird never succeeded in playing recognisable pictures back from his Phonovision discs.

Baird's experiments became obsolete with the rise of electronic recording, which synchronised images with sound recordings on magnetic tape. The US company Bing Crosby Enterprises first demonstrated video recording in 1951, and by 1953 the first video recorder had been built by RCA, but the playback quality in both of them was very poor.

The Californian company Ampex produced the first commercial video recorder: the VR 1000, over twice the size of a wardrobe, was launched at a broadcasting exhibition in Las Vegas in 1956. It ran on tapes 5cm (2in) wide and 800m (875yd) long and was initially intended for use by television stations but soon led to the development of home video recorders. Telcan, the first video recorder for the domestic market, also used reels of tape and was first sold in Britain in 1963 by the Nottingham Electronic Valve Company.

MANY POPULAR PROGRAMME FORMATS HAD EARLY ORIGINS. *SPELLING BEE*, THE FIRST GAME SHOW, WAS SCREENED BY THE BBC IN 1938 WHILE *FARAWAY HILL*, THE FIRST TELEVISION SOAP OPERA, WAS SHOWN IN WASHINGTON AND NEW YORK FOR 30 MINUTES EVERY WEDNESDAY FROM 1946 ONWARDS. THE WORLD'S INAUGURAL COMMERCIAL TELEVISION STATION, WNBT NEW YORK, OPENED IN 1941.

Fairs and carnivals
Entertaining the crowds

All over medieval Europe people made merry at the fair. Acrobats, stilt walkers and jesters entertained the crowds, and trinkets could be won as prizes. But although most of these carnivals coincided with religious occasions, fairs evolved from livestock sales and the annual hiring of labourers. Here craftsmen and merchants sold their wares, as the Romans had done at the trade fairs they established at business venues.

A merry-go-round operated in 1620 in Turkey, but roundabouts had existed in Byzantium in the 5th century. In Britain a hand-powered carousel was set up in London in 1729 at Bartholomew Fair, a cloth fair instigated in 1120 by Rahere, the jester of Henry I. Big wheels appeared from the 15th century. They became known as Ferris wheels when George Ferris, the American owner of a company that produced big wheels, built a huge steam-powered version that could hold more than 2,000 people for the World's Columbian Exposition in Chicago, in 1893.

Taking holidays
Relaxation and celebration

The first holidays were 'holy days', set aside for religious devotion: the Egyptian pharaohs forbade work on about 70 days each year. In ancient Rome more than 100 feast days were dedicated to gods and goddesses. Work days were known as *dies vacantes*, or vacant days, from which, through a reversal of meaning, 'vacation' derives. Both the Jewish Sabbath and the Christian Sunday began as religious rest days based on the Biblical Creation story in which God 'blessed the seventh day' and 'rested from all His work'.

The early Church absorbed pagan feast days into its calendar. In the Middle Ages working on a public holiday without official dispensation was a punishable crime equal to murder. The celebration of saints' days lapsed in England and other Protestant countries following the Reformation of the 16th century.

When Europe began to change from an agrarian, rural society to an industrial, urban one in the late 18th century, manual workers were granted little time away from their jobs. The first employees in Britain to benefit in the long struggle for paid leave were workers at the South Metropolitan Gas Company in London. In 1872 they were granted annual leave with pay 'as a mark of appreciation of their conduct'.

ON VACATION

✣ William Wordsworth and other Victorian Romantic poets fostered interest in the natural beauty of such areas as the Lake District and Scotland, which became holiday destinations.

✣ Rules were strict at Joseph Cunningham's Young Men's Holiday Camp, which opened in 1894 in Douglas, Isle of Man. Alcohol and gambling were forbidden and holidaymakers were fined for returning to the camp after 11.45pm.

✣ The principle behind dodgems, using a metal floor and electrically charged ceiling to power 'cars', was patented in 1890 by a New Yorker named James Adair, but the idea was not realised until the 1920s, in rides called Gadabouts.

✣ The South African-born Billy Butlin opened his first venture in 1936 near Skegness, Lincolnshire. Facilities at the 600-chalet holiday village included a theatre, recreation hall, swimming pool and tennis courts.

Spa and seaside
Taking the waters

The West Country town of Bath was reputedly founded on the spot where Bladud, father of King Lear, was cured of leprosy in 836 BC by immersing himself in steaming swamps. The Romans appreciated Bath's waters and drank them for their health-giving properties. In the Middle Ages pilgrims would visit Bath and other such European sites to be healed. In the 16th century watering-places, including Vichy in France and Baden-Baden in Germany, became popular resorts and one of the most stylish was the Belgian town of Spa.

An appreciation of the merits of the seaside began in the 1730s in Scarborough in Yorkshire, a town known for its medicinal springs. Sea bathing, regarded as shocking by many, was sanctioned by the medical profession, including the physician Richard Russell. He recommended that sufferers of all complaints, from gonorrhoea to gout, drank seawater mixed with milk or port.

Advances in engineering combined with the Victorian love of promenading resulted in the conversion of the boat jetty to the seaside pleasure pier. The first was effectively a large suspension bridge, designed by Captain Samuel Brown, a naval architect, for Brighton. The first iron pier, with a frame sunk into the seabed, was constructed at Herne Bay in Kent in the 1860s.

IN 1871 THE BANKER AND LIBERAL MP SIR JOHN LUBBOCK HELPED TO PASS THE BANK HOLIDAYS ACT. IT OBLIGED ALL BANKS IN ENGLAND AND WALES TO CLOSE ON CHRISTMAS DAY AND BOXING DAY, GOOD FRIDAY, EASTER MONDAY, WHIT MONDAY AND ALSO ON THE FIRST MONDAY IN AUGUST.

The hotel
A room for the night

Accommodation for travellers sprang up along ancient trading routes. In ancient Persia inns, which were usually built around a central courtyard, were known as caravanserais. During the early Middle Ages travellers lodged at monasteries. From the 11th century the advent of the Crusades and pilgrimages, as well as the development of trade, meant increasing travel and the opening of many lodging houses. These short-term resting places offered few facilities; guests often had to share rooms, or sometimes even beds.

The oldest known hotel, the Hōshi Ryokan, was founded in Japan in AD 717. The Hotel, named from *hôtel*, the French for a large private mansion, opened in Exeter, Devon, in 1770 (it is now the Royal Clarence). David Low began business at his Grand Hotel in London in 1774. Low had been a hairdresser, but most early proprietors had a background of serving the aristocracy.

Holidays abroad

'We would venture anywhere with such a guide and guardian as Mr Cook, for there was not one of his party but felt perfectly safe when under his care.'

MATHILDA LINCOLNE AND HER SISTERS, *THE EXCURSIONIST*
(THOMAS COOK'S TRAVEL NEWSPAPER), 1855

Until Christian pilgrims began to visit Jerusalem from AD 66 to enrich their faith, the purpose of foreign travel was to explore, conquer and trade with neighbouring lands. Money also made possible the 'Grand Tour' of the Continent from the mid 16th century. The sons of British aristocrats, who were the first people to be described as 'tourists', were sent on journeys around the capitals of Europe. Travelling in the charge of a tutor, the purpose was to complete their education.

Journeys for pleasure, which were affordable to the new industrial magnates as well as the aristocracy, burgeoned in the 19th century. The pioneer of the package holiday, with both travel and accommodation provided, was a Mr B Emery, the son of a Swiss watchmaker who had settled in England. Following the end of the Napoleonic Wars, from about 1816, Emery arranged for groups of up to six people to travel from London to Switzerland by stagecoach on 16-day trips costing 20 guineas.

Tourism begins

The first tourist travel agency was set up by Thomas Cook, a temperance advocate and wood-turner, who ran his first excursion, a day trip from Leicester to a temperance meeting in Loughborough, in 1841. The cost of the journey was 1 shilling (5p). By the time Cook opened a permanent office in London in 1865, his horizons had broadened considerably. A decade earlier he had led a trip to Paris via Cologne. The highlight was an excursion down the Rhine re-enacting the journeys that had inspired poets such as Lord Byron.

Travel by sea was an integral part of Thomas Cook's long-haul itineraries. From the late 19th century passengers travelling to India aboard the liners of the Peninsular and Oriental Steam Navigation Company (P&O) discovered that the

best cabins were those protected from the sun's glare. Wealthy travellers became accustomed to asking for a 'posh' ticket: port (side) out, starboard home. However, such seaborne holidays only became fashionable after 1918 when redundant warships were put to enterprising use.

The Greek writer Philostratos recommended in about AD 170 that 'those of advanced years should bask in the warm sun'. But to the Victorian and Edwardian tourists who visited the Mediterranean to enjoy its mild winters, pale skin was de rigueur. The vogue for sunbathing is thought to have started with American expatriates Gerald and Sara Murphy. After the 1922 winter season they persuaded the owner of the Grand Hôtel du Cap at Antibes on the French Riviera to remain open. From here they ventured each day to sunbathe on the beach. Two years later the Murphys bought a villa in Antibes and invited such friends as the American writer F Scott Fitzgerald to join them: the Riviera soon became a playground for the rich.

Travel writers

Herodotus, the 5th-century BC Greek historian and voyager, was the first author to describe the places he visited. He wrote of the Babylonians who had 'long hair and cover themselves with perfume' and marvelled at 'beasts in Libya that go without water'. The oldest known printed guidebook, Benedict's *The Wonders of Rome* of 1473, advised on notable sites to visit.

The Grand Tourists had hired guides to conduct them around Europe, but the 19th-century sightseer needed a cheaper alternative. Publisher John Murray's first guidebook for English tourists abroad, *Travels on the Continent*, appeared in 1820. In Germany a guide to Koblenz was compiled in 1829 by Karl Baedeker. A decade later his guides included 'stars' awarded to outstanding hotels, restaurants and attractions. The idea was so popular that all guides became known as 'Baedekers'.

To afford travellers safe passage the Egyptian pharaohs issued them with cartouches, oval frames enclosing an engraving of their ruler's name. In the 2nd century BC the Greeks used 'letters of confidence' as early 'passports'.

Playing dice
The lucky throw

Dice-playing was so popular in ancient Greece that triple six, the best throw of all, became synonymous with good fortune. The Romans too were keen players and even had professional associations for those who made their living from gambling. The emperor Claudius wrote a book on the subject of playing dice in the 1st century AD.

Such discoveries as a set of bone dice and the box in which they were stored, found in Glastonbury, Somerset, show that playing dice also spread through Iron Age Europe. By this time dice had been around for thousands of years – they have been found in 5,000-year-old Sumerian graves and in the ruins of cities in India and Pakistan built in around 2300 BC.

In western Turkey the Lydians used dice to amuse themselves from about 1300 BC. Herodotus, the Greek historian, reported in the 5th century BC that during a long famine King Atys of Lydia allowed his subjects to eat only on alternate days. On fasting days 'the game of dice, the game of knuckles, games of ball and other games were invented' to help to pass the time.

FRAUD AND FORTUNE

✤ Playing for money has encouraged deception from early times: digs at Roman sites have unearthed loaded dice weighted with spots of mercury.

✤ In Roman times dice-playing was illegal outside the midwinter festival of Saturnalia but soldiers took dice on campaigns for all-year amusement.

✤ Premium Bonds, in which investors forgo interest in exchange for a chance of monthly cash prizes, were launched by the British Government. ERNIE (the Electronic Random Number Indicator Equipment) drew the first winning numbers in 1957.

Gambling on sport
The bookies and the pools

Private wagers on two-horse races were commonplace among the aristocracy when horse-racing became a popular sport in medieval times. At the racetrack betting against odds set by a bookmaker began in the 19th century. Such gambling remained 'on course' until 1886 when a Harry Schwind and a Mr Pennington set up a partnership at Ladbroke Hall in Worcestershire to act as 'Commission Agents and Bookmakers to the Aristocracy'. Odds were available to punters to bet on the horses that Schwind trained there. Six years later Ladbrokes was bought by the entrepreneur Arthur Bendir, who moved it to London. Guards officers were employed as commission agents and they helped to ensure Bendir's success in cultivating the rich and famous. From 1913 clerks took bets at small curtained-off desks, ensuring total discretion.

John Barnard, a former officer in the Coldstream Guards, set up Pari-Mutual Pools in 1922, issuing coupons on which punters could bet on the outcomes of six football matches. In the first year of operating Barnard could not even cover his postage costs, but his idea gradually found favour and the coupons started flooding back during the following year.

Inspired by Barnard, John Moores, a telegraphist at the Commercial Cable Company's Liverpool office, started a rival operation with two of his colleagues in 1923. To keep their identities secret from their employers the trio printed coupons under the name 'Littlewood', the original surname of one of them, Harry Askham. Starting with capital of £150, and hiring boys to distribute the first 4,000 coupons outside Manchester United's stadium, their first season's accounts showed a loss of £600. Two of the partners lost their nerve and withdrew, but Moores persisted and by 1930 he was a millionaire.

The casino
Lady Luck's palace

In 1713 the first recorded gambling establishment was opened at the Duc de Luxembourg's Paris home, the Hôtel du Perron, by Ferencz Rákóczy II, Prince of Hungary and the exiled ruler of Transylvania. Drunken brawls were common and winners could spend their money on the resident prostitutes.

The first legal casino, the Promenade House at Baden-Baden, in south-west Germany, opened in 1765. The Flamingo was the first of the grand casino-hotels for which Las Vegas, Nevada, is famed. Built in 1946 by the gangster 'Bugsy' Siegel, it lost $100,000 in the first two weeks of operating. When Bugsy was killed in 1947 another gangster, Gus Greenbaum, took over the casino. He made a profit of $4 million in his first year alone, which encouraged other crime syndicates to establish casinos in the town.

The invention of roulette is often credited to Blaise Pascal, the 17th-century French mathematician who developed many devices to illustrate chance. But it is more likely that the roulette wheel derives from the 18th-century game of hoca, in which a ball is spun into a circular plate that contains 40 pockets around its edge.

In 1890 a British patent was issued for a mechanism with multiple dials and a window for viewing symbols. The design was adopted in 1895 by the Bavarian-born Charles Fey of San Francisco whose 'Liberty Bell' was the world's first one-armed bandit. The alternative name 'fruit machine' derives from 1908, when a gambling ban in some states led to fruit gum being given as prizes rather than cash.

FOR 5,000 YEARS GAMESTERS HAVE PLAYED WITH DICE MADE OF WOOD, STONE OR BONE. IN ANCIENT GREECE ANIMAL KNUCKLEBONES WERE USED IN SIMPLE GAMES OR ADAPTED INTO RUDIMENTARY DICE. VALUES WERE ASSIGNED TO FOUR SIDES OF THE UNEVEN, OBLONG BONES, WITH NUMBERS OF OPPOSITE FACES ADDING UP TO SEVEN.

Circus acts
Travelling entertainers

The original circus was the racetrack of Roman times, with long straight sides and semicircular ends, but today's version was devised in England by a former cavalry officer, Philip Astley, in 1768. It gained the name 'circus' in 1782, when rival horseman Charles Hughes opened his Royal Circus in London. By 1777 Astley had created a show that incorporated 'The Little Military Learned Horse', a strongman named Signor Colpi, clowns and acrobats. A monkey was soon added, and in 1832 a lion, a tiger and some zebras were led around the ring.

Acrobats, jugglers and clowns all performed in ancient times, as did animal tamers. Alfred the Great, the 9th-century king of Wessex, was once entertained by a wild-beast show. But the first of the modern animal tamers, the Frenchman Henri Martin, performed with lions, an elephant and a boa constrictor in 1831 at the Cirque Olympique in Paris. Morok the Beast Tamer, born the American Isaac Van Amburgh, claimed to be the first person to put his head inside a lion's mouth. In 1838 he first presented his act in the ring of Astley's Amphitheatre.

Art galleries and museums
Treasures on display

THE POPULAR CIRCUS ACT, THE FLYING TRAPEZE, WAS BORN OUT OF THE DARING OF THE FRENCHMAN JULES LÉOTARD, WHO FIRST PERFORMED WITH THE CIRQUE NAPOLÉON IN PARIS IN 1859 AT THE AGE OF 21. INITIALLY A PILE OF MATTRESSES WAS THE ONLY PROTECTION AGAINST DANGEROUS FALLS FOR LÉOTARD AND OTHER CIRCUS PERFORMERS. THE SAFETY NET WAS INTRODUCED BY A SPANISH ACROBATIC TROUPE, THE RIZARELLIS, AT THE HOLBORN EMPIRE IN LONDON IN 1871.

As proof of their status the rich and powerful of ancient Egypt, Babylonia, China and India amassed precious objects. And although the Greeks were probably the first to collect and display art for its beauty, it was the Romans, by systematically plundering the nations they colonised, who stimulated a wider interest in artistic accomplishments.

In medieval Europe art was not collected other than by churches and monasteries, and it was only during the 15th century that noble families such as the Medicis in Italy began buying art and displaying it in their own homes. Private collections of unusual items were also being put together at this time, when they were known as 'cabinets' from the Italian *gabinetto*, meaning 'cage' or 'basket'. In 1523, as the beneficiary of the brothers Domenico and Antonio Grimani, the Venetian Republic became the first public body to inherit and exhibit such a cabinet. A similar bequest to Oxford University by the explorer and collector John Tradescant led to the opening in 1683 of the Ashmolean Museum. On view were exotic objects ranging from stuffed animals and birds (including a dodo, a flightless bird from Mauritius that would be extinct by the end of the century) to weapons, armour and coins.

The British Museum, which grew from several cabinets, including that of the physician and explorer Sir Hans Sloane, was the first national collection. After Sloane's death in 1753 the British Government accepted responsibility for maintaining the 100,000 paintings, as well as for the books and specimens. In 1759 the museum was opened in Montagu House, Bloomsbury, and admission was free. The museum was rebuilt on the same site between 1823 and 1852.

In 1776 the Earl of Pembroke opened his home, Wilton House near Salisbury, to the public free of charge. The idea was immediately popular and that year the house received some 2,500 visitors. When the Earl of Leicester granted access to Holkham Hall in Norfolk a year later, 'noblemen and foreigners' were allowed to visit from Mondays to Saturdays. 'Other people' were allowed on Tuesdays only. But it was financial necessity, not generosity, that led the 6th Marquis of Bath to open Longleat in Wiltshire to the public in 1949, making it the first stately home to be run as a commercial enterprise.

Botanical gardens and zoos
Wonders of the natural world

In the early 15th century BC Queen Hatshepsut of Egypt imported frankincense trees and other exotic plants from Somaliland, which she displayed with wild animals. But it was her successor, Pharaoh Thutmose III, who extended the botanical collection with plants and seeds from Palestine and Syria.

Pleasure parks for entertainment, planted with ornamental trees, were built by both the Greeks and Romans, and similar parks existed in India by the 3rd century BC. The first modern European botanical garden, founded in Pisa, Italy, in 1543, was greatly influenced by contemporary discoveries in the New World, such as the Aztec gardens of Montezuma I at Huaxtepec in Mexico.

Solomon, the Israelite king of the 10th century BC, kept menageries. In the 4th century BC the 300 creatures collected by Alexander the Great became the subjects of Aristotle's *History of Animals*. When the Roman Empire fell, zoos disappeared and were not revived in Europe until the Middle Ages.

Henry I of England put the animals presented to him by other monarchs in a small zoo at Woodstock, Oxford, around 1100. In the 13th century three leopards given by Frederick II of Sicily to his brother-in-law Henry III became the first residents of London's Tower Menagerie. They were joined by a 'white bear', a polar bear from Norway, which was regularly observed fishing in the Thames.

The French king's menagerie at Versailles, where animals were kept for research and education, became a modern zoo in 1665. In London the royal tradition of housing wild animals at the Tower continued until 1828, when all the creatures were moved to the Zoological Gardens in Regent's Park under the management of the Zoological Society of London. Here the world's first reptile house opened in 1849, and the first insect house in 1889.

THE GREAT EXHIBITION

It was Prince Albert, Queen Victoria's husband, who suggested boosting British industry by inviting exhibitors to display their achievements, and in 1851 'the Great Exhibition of the Works of Industry' opened in Hyde Park, London. The event was housed in a huge structure of glass and iron designed by Joseph Paxton and dubbed by *Punch* magazine 'the Crystal Palace'. The exhibition, described by Victoria as 'magical ... so vast, so glorious, so touching', provided the nucleus of the collections for both the Victoria and Albert Museum, founded in 1852, and the Science Museum, which opened in 1857.

The Olympic Games
Sporting ideals

Athletic sports were popular among the ancient Egyptians, but in Greece sporting ability also took on a religious dimension. At certain sacred sites dedicated to the gods events grew in prominence, drawing athletes and spectators from the various city-states. Most renowned of these were the Pythian Games at Delphi; the Nemean Games held at Nemea or Argos; the Isthmian Games at Corinth; and the Olympic Games, held in honour of Zeus, the father of the gods, every four years at Olympia. All four events comprised the Crown Games, named from the crowns of leaves given to winners.

The original Olympic victory was recorded in 776 BC when Coroebus of Elis, a cook, won the *stadion*, a track race of approximately 185m (200yd), which gave rise to the word 'stadium'. According to legend, this was the furthest that the hero Herakles (known to the Romans as Hercules) could run in a single breath.

By 500 BC the games were attracting some 40,000 spectators. As part of the religious ritual athletes, accompanied by their trainers, would arrive a month in advance for a strictly supervised pre-games ritual. From the beginning contenders competed for money and bribery was commonplace. The Roman emperor Theodosius, who banned all pagan festivals, ended the games in AD 394.

MODERN MOVEMENT

In Britain the Cotswold Olympics were held in 1604, nearly 300 years before the modern Olympics began. Between 1859 and 1889 the Greeks tried to reintroduce the games. But their efforts were thwarted by rioters protesting that only wealthy athletes could afford to travel to Athens to train. At about this time Baron Pierre de Coubertin, a French educationalist, suggested to his country's government that it should find a way to improve the physical condition of France's youth.

The Olympic flame, which burns constantly throughout the games, was first used at Amsterdam in 1928. It may recall relay races run in ancient times (although not at the games in Olympia) which used torches, not batons.

Coubertin's plan for reviving the Olympic Games, not just for the French but for young people from other lands, was accepted, and the International Olympic Committee was formed in 1894. It resolved that 'sports competitions should be held every fourth year on the lines of the Greek Olympic Games'. The modern Olympic Games were inaugurated in Athens in April 1896. The first winter games were held at Chamonix, France, in 1924.

Body contact sports
Wrestling, boxing and the martial arts

Throws, headlocks, strangleholds and other skills of unarmed combat gave rise to wrestling, a pastime depicted in Egyptian wall paintings at Beni-Hasan in around 3400 BC. The Babylonians, Assyrians and ancient Chinese also wrestled, but it was the ancient Greeks who made the sport their own, crediting the mythical hero Theseus, who slew the Minotaur, with having laid down the first rules of 'scientific wrestling' in about 900 BC.

Greek wrestlers oiled their bodies. They then sprinkled powder or sand over the oil to provide a 'second skin' and to help the body to stay warm. Wrestling was an Olympic sport from 704 BC. Its most fearsome and often fatal version was the *pankration*, introduced in 648 BC, which combined elements of boxing and wrestling. Only biting and gouging were forbidden.

Hieroglyphs drawn by the Egyptians in the 4th millennium BC suggest that boxing reached them from Ethiopia, and then spread north to Greece. From 688 BC boxing was included in the Olympics, but despite its violence deaths were rare. The sport, which was adopted by the Romans, was later revived in 17th-century England. The first recorded bare-knuckle fight took place in 1681 between a butcher and the Duke of Albemarle's footman. In 1719 James Figg, who had won so many fights that he was acclaimed as the champion of England, set up the first modern school of boxing in London. On his business card he described himself as 'Master of the Noble Science of Defence'.

COMPETE TO WIN

✷ Leonidas of Rhodes won the foot-race in four Olympiads between 164 and 152 BC.

✷ In 1812 sportsmen competed at the first modern athletics contest, held at the Royal Military College, Sandhurst, England.

✷ Judo, whose name means 'easy way', was developed in Japan in the 19th century. It derived from the ancient martial art of jujitsu, which had fallen into disrepute by this time because of its dangerous techniques. Judo's inventor was Dr Kanō Jigorō, who opened his first dojo, or 'training hall', in 1882.

Athletic competition
The earliest events

Although by 3800 BC organised races probably took place in Egypt, the Greeks were the masters of athletics. The *daiulos*, a race to the end of the track and back, was added to the one-length *stadion* in the Olympics of 724 BC. It was followed four years later by a *dolichos* or endurance race. In 708 BC the pentathlon, comprising the five disciplines of long jump, discus, javelin, running and wrestling, was introduced. Long jumpers often held weights to provide extra momentum.

After the demise of the ancient games, track and field sports were rare until the 12th century. The marathon, a 42.195km (26 mile) race, dates back to the 1896 Olympics. It commemorates Pheidippides, who ran from Athens to Sparta with a plea for help after the Persians, en route to attack Athens, reached Marathon in 490 BC.

The rise of football
The beautiful game

Four thousand years ago the Egyptians were playing catch or casual kicking games with balls made of leather or linen and stuffed with reeds or straw. A kind of football may have been played in ancient China around 200 BC. Here *t'su chu* – literally, 'to kick a stuffed leather ball' – was enjoyed by players who attempted to kick the ball through a hole in a silk net.

Two teams were needed for *episkuros*, a popular ball game in ancient Greece and later in Rome, where it was known as *harpastum* or *pila pedalis*.

One Roman stone tablet from the 2nd century AD and discovered at Sinj in modern-day Croatia shows a man holding a football made up of hexagonal sections.

The Roman legions may well have brought football to northern Europe; certainly by the 12th century a lawless, violent type of football was common in Britain. The ball was an inflated pig's bladder encased in leather. Shrove Tuesday was a traditional day for football, and in such Roman cities as Chester the game was said to have been played to commemorate the Romans' departure.

Feuds between football teams and their fans concerned medieval monarchs, starting with Edward II in 1314 who proclaimed: 'Forasmuch as there is a great noise in the city caused by hustling over large balls, from which many evils may arise, which God forbid, we command and forbid, on behalf of the King, on pain of imprisonment, such game…' The name of the game first appeared in an Act passed in 1424 forbidding the game of 'fute ball'.

FIRST ELEVEN

In the 17th century a violent version of football was still being played at schools such as Eton and Winchester and only in 1848 did Cambridge University attempt to standardise rules for a safer eleven-a-side game.

The Football Association (FA) was set up in 1863, when representatives of clubs and schools began the task of establishing a set of rules. By 1871 FA membership consisted of 30 clubs from London and the Home Counties. The following year the first FA Cup final was held at the Oval cricket ground, London: Wanderers beat the Royal Engineers 1–0. The desire for regular matches and payment for players led a dozen clubs to form the Football League in 1888. The word 'soccer' is derived from 1890s' Oxford University student slang for 'Association'.

Football was a brutal game in the 16th century. According to Puritan Philip Stubbes, in his *Anatomie of Abuses* of 1583, it 'causeth fighting, brawling, contention, quarrel picking, murder, homicide and great effusion of bloode'.

Playing rugby

Running with the ball

In 1823, so sporting legend relates, William Webb Ellis, a student at Rugby School, caught the ball during a game of football and ran with it. In other versions of the traditional tale, the first handling of the ball is dated to 1838 or 1839 and the instigator is said to have been a boy named Mackie.

Whatever the truth, from 1841 'running-in' with the ball was permitted at Rugby and a few other public schools. In 1846 *The Laws of Football as Played at Rugby School* were published, establishing rules for scrummaging. It also stated that 'All matches are drawn after 5 days or after 3 days if no goal has been kicked.' Points were not scored for a touchdown, but for a subsequently successful 'try' at kicking the ball over the crossbar.

In 1871 the Rugby Football Union (RFU) was created as the governing body of the sport in England. The size of teams remained fluid until 1875, when sides of 15 players were introduced in a match between the universities of Cambridge and Oxford.

A dispute in 1895 over financial compensation for rugby players' unpaid time spent away from their jobs led 22 clubs from the north of England to form the breakaway Northern Football Union. The NFU developed its own rules, and in 1906 stipulated that sides should consist of only 13 players. It renamed itself the Rugby Football League in 1922.

SPORTING MILESTONES

- In 1497 a football was bought for James IV of Scotland: its price was 2 shillings (10p).
- Charles II attended a football match between his servants and those of the Duke of Albemarle in 1681.
- The first known golfing trophy was awarded in Edinburgh in 1744.
- In 1764, the golf course at St Andrews, Scotland, became the first with 18 holes.
- England played Scotland at rugby for the first time in 1871.
- In 1875 the first known football programme was prepared for a match at Hampden Park between Queen's Park and Wanderers.
- Molesey Ladies, the first women's hockey club, was founded 1887.

A round of golf

The game of clubs

Many cultures boast games similar to golf. In the 10th-century Chinese game *chiuwan*, bamboo-shafted clubs were used to hit hardwood balls into holes in the ground marked with flags. Traders may have brought the game to the West, but similar European games such as the Dutch *kolven* or *kolf*, in which a ball was aimed at posts placed at each end of a short course, probably arose independently from the Middle Ages onwards.

The object of the game that was played from the 15th century on the east coast of Scotland, from which modern golf originates, was to hit a ball into a distant hole. By 1457 the game was well established and was soon taken up by women. Mary, Queen of Scots was censured by the Church in 1568 for playing golf so soon after the murder of her husband, Lord Darnley, the previous year.

Early courses were rough, with narrow fairways, but from the middle of the 19th century they gradually began to take on their modern look with clipped greens. The employment of 'caddies', from the Old French *cadet*, a page at court, to carry golfers' clubs dates from the 1790s.

Real and lawn tennis
Game, set and match

A French handball game, *jeu de paume* or 'game of the palm', played from the 12th century in French monastery cloisters, became so popular that Paris had at least 13 manufacturers of balls for the game by 1292. Its modern name, tennis, comes from the French *tenez*, a word probably called out by the server before hitting the ball and shortened from *tenez-vous prêt*, or 'get ready'. The game was first recorded in English as 'tenetz' by the poet John Gower in 1399.

Early players hit the ball with their bare hands but soon devised large gloves and eventually, to make them lighter, the centres were replaced with a taut, elastic rope weave. By about 1500, short-handled rackets appeared, and both hands and rackets were used to hit the ball.

Tennis was popular with the French kings, who did much to introduce the game to the rest of Europe. When Henry V received some balls from the future Charles VII of France in 1414, he started a tradition of aristocratic tennis playing in Britain. The first tennis world championships, the oldest of all sporting world championships, was won by the French player Monsieur Clergé in about 1740. He held his title for nearly a decade.

The game at which Clergé excelled was what we now know as real tennis. Played in an enclosed court, the ball is allowed to bounce off a wall. The modern game of tennis is a direct descendant of real tennis. The inventions of the rubber ball, which could bounce on grass, and the lawn mower, used to create a smooth playing surface, meant that tennis

---

BAT AND BALL

✤ Baseball is first mentioned in *A Pretty Little Pocket Book*, an English book of games published in 1744.

✤ Squash was probably invented by boys at Harrow School, and was being regularly played there by 1850.

✤ In about 1879 the Duke of Beaufort adapted the ancient children's game of battledore and shuttlecock, playing it over a net at Badminton, his Gloucestershire home, from where the game drew its new name.

IN 1874 MAJOR WALTER CLOPTON WINGFIELD OF WALES PUBLISHED RULES FOR LAWN TENNIS PLAYED ON AN HOURGLASS-SHAPED COURT. THE FOLLOWING YEAR HE TOOK OUT A PATENT FOR WHAT HE CALLED 'SPHAIRISTIKÈ' (FROM A GREEK WORD FOR 'BALL GAME'). THE GAME, HE EXPLAINED, COULD 'BE PLAYED IN ANY WEATHER BY PEOPLE OF ANY AGE AND BOTH SEXES'.

began to be played regularly on grass in the mid 19th century. More powerful rackets with longer handles appeared at about the same time. The first lawn tennis club was founded in 1872, at Leamington, Warwickshire.

In 1875 J M Heathcote, a real tennis player who had started to play lawn tennis, asked his wife to cover a ball in flannel to make it easier to hit on wet grass, and the idea quickly caught on. Lawn tennis became so popular that the All-England Croquet Club at Wimbledon, in London, was persuaded to set aside a lawn for tennis. In 1877 it became the All-England Croquet and Lawn Tennis Club and as such it instituted the first Wimbledon Tennis Championships, in which a rectangular grass court was used. Only 22 players, all of them men, entered the championship, which was won by Spencer Gore. The first winner of the women's championship, inaugurated in 1884, was Maud Watson.

The game of cricket
Leather on willow

No one knows when countrymen of the Weald in southern England began using stones and sticks to play the game destined to become cricket. What is certain is that a sport developed from a shepherds' custom: one stood in front of a wicket-gate with a stick, while another hurled a stone to hit the 'bail', or crosspiece, on the gate. 'Bail' is Old French in origin, going back to *baillier*, 'to enclose', or even further back to the Latin *baculum*, 'stick'.

'Cricket' probably comes from *crok*, an Old English word for a shepherd's crook or curved staff. Another possible early name for the game was 'creag'. In 1300 the royal wardrobe accounts recorded 100 shillings being laid out for Prince Edward to 'play at creag'. The word 'cricket' was in use by 1550, when, as 'a scholler in the free school of Guldeford', John Derrick, later to become a coroner, 'and several of his fellowes did runne and play there at crickett and other plaies.'

The laws of the game were laid down in 1744. Cricket was becoming widely popular, particularly in southern England, by the time that Thomas Lord, the Yorkshire-born entrepreneur and a keen cricketer, opened his ground in 1787. It was in Dorset Fields (now Dorset Square), part of the Marylebone estate.

Lord had been attached to the White Conduit Club where he met the Earl of Winchelsea. The earl and other aristocrats encouraged Lord in his venture, and the Marylebone Cricket Club (MCC), which became responsible for the laws of the game, was born. Lord's ground and the MCC moved to their present location in north-west London in 1814.

THE TEST MATCH

Cricket travelled all over the world with the British Empire, and was introduced into Australia by a ship's crew in 1803. In 1877 Australia beat England in the first Test (international) match, held in Melbourne. After an 1882 match, played at the Oval in London, in which England lost again to Australia, the *Sporting Times* commented that English cricket would be 'cremated and its ashes taken to Australia'. This led to the Ashes games between Australia and England, whose trophy is an urn (which remains permanently at Lord's) containing the ashes, probably of a bail burned on England's tour of Australia in 1882–83.

On horseback

'His colour is Bay, and his near foot before with both his hind
feet have white upon them, he has a blaze downe his face,
something of the largest. He is about 15 hands high, of the
most esteemed race among the Arabs.'

THOMAS DARLEY, WRITING ABOUT THE DARLEY ARABIAN COLT, 1703

In the steppes, those grasslands that extend from the Ukraine to Mongolia,
horses were first corralled for food in around 4300 BC. It was here, too,
where horses had roamed wild for centuries, that someone was first
inspired to jump up on horseback. This lasting partnership predated the wheel
as a form of transport by about 500 years.

Riding spread through both trade and war, but migration was slow and
horse-drawn chariots did not appear in the Middle East until about 1800 BC.
Horses made their sporting debut at the Olympic Games in Greece where four-
in-hand chariot races were run from 680 BC.

Early riders sat bareback or on blankets thrown over horses' backs, but
saddles made balancing much easier. Leather saddles were developed by the
Scythians, nomadic warriors of central Asia, between the 3rd century BC and
the 1st century AD. The saddle was further improved in medieval Europe to
hold heavily armed knights more securely in place.

Stirrups and shoes

Scythian horsemen also devised the first stirrups, pictured on a vase of the 4th
century BC from Chertomlyk in the Ukraine, and the horse collar. Scythian
stirrups of iron are said to have been introduced to Europe by the
Asiatic chieftain Attila the Hun, who repeatedly invaded the Roman
Empire in the mid 5th century AD. Such stirrups were certainly
being used in China and Japan by AD 600.

Horseshoes were fitted in Eurasia by the 2nd
century BC. Nailed iron horseshoes probably
arrived in Europe around the 5th
century AD, introduced by invaders
from the east, although they may
also have been independently invented
by the Romans.

Bareback horse-riding became an Olympic
event as early as 648 BC, but the modern
sport of racing with horses was
developed by the Arabs. Horses had
already been raced long before the first
hippodrome, or racing track, was built in Baghdad
in the 8th century AD.

When the Crusaders returned to Europe between the 11th and 13th centuries they brought Middle Eastern horses with them. These were raced as a way of showing off their speed to prospective buyers. The secretary to Thomas Becket, the Archbishop of Canterbury, wrote in about 1160 that 'jockies, inspired with thoughts of applause, and in the hope of victory, clap spurs to the willing horses, brandish their whips, and cheer them with their cries.' The first record of a financial prize occurs some 30 years later, when £40 was offered by Richard I to the winner of a 4.8km (3 mile) race contested by knights.

The sport of kings

In the early 15th century races were initiated in Europe at Sanlúcar de Barrameda in Spain. Britain's most ancient race is most probably the Kiplingcotes Derby, at Market Weighton, Yorkshire, which has been run on every third Thursday in March since its inception in 1519.

Racing in Britain became more organised in the 17th century when Charles II established spring and autumn meetings at Newmarket, Suffolk, which remain pivotal to the racing calendar today. For many of the races, the king donated prizes of 100 guineas' worth of silver plate. Two or sometimes three horses would take part in each race, probably because no more were available locally. To win, a rider had to be victorious in two heats. For 'sweepstakes' the prize was a purse to which each owner had contributed beforehand.

As racing became more fashionable, fields grew larger and new courses were established, including Royal Ascot, founded by Queen Anne in 1711. Modern racing began with the inauguration of the English Classics – the St Leger in 1776, the Oaks in 1779 and the Derby in 1780.

Steeplechases, run on a racecourse provided with artifical obstacles over which a horse must jump, probably originated in Ireland during the 18th century. Riders would race their steeds across the rough terrain of the countryside; local church steeples acted as landmarks along the course.

The first woman jockey on record is 22-year-old Alicia Meynell, the mistress of a Colonel Thornton. In 1804 Meynell rode sidesaddle on her lover's 20-year-old horse Vingarillio against a Captain William Flint, over a 6.4km (4 mile) course at York. Meynell started as favourite at 5–4 on, but lost the race.

Skiing and skating
Sports on snow and ice

The god of winter, so Norse legend relates, walked the world on skis. Pine and spruce skis 5,000 years old have been found in Scandinavian bogs, and skiers depicted 4,000 years ago in Norwegian rock paintings resemble their modern counterparts. In such sub-Arctic areas as Siberia short, wide snow boards were worn, sometimes with animal pelts attached to their undersides to prevent dangerous slipping.

Skis began as curved frames covered with leather, and lengthened in northern Europe; examples about 4,000 years old found at Kalvtrask, Sweden, measured more than 2m (6ft).

By the 15th century most of the armies in northern Europe were using skis. Skiing was recorded in parts of Slovenia in the late 17th century, but was unknown in France, Italy or Switzerland until it was introduced by British and Scandinavian travellers in the 19th century.

Downhill and competitive skiing started after 1840, when a Norwegian named Søndre Norheim replaced the usual loose leather strap (which made sharp turns impossible) with a firm binding made from birch roots soaked in hot water. This binding held the heel and instep of the boot to the ski.

The first skiing races were held near Oslo, Norway, in 1866 and by the turn of the century competition was widespread. International downhill slalom racing, named from the Norwegian *slad låm*, or 'sloping path', began in 1927, but piste skiing still involved arduous climbs. In the Swiss Alps trains used by summer visitors were adapted for skiers, and in 1932 the Parsenn cable car opened at Klosters in Switzerland.

WINTER LEISURE

Two thousand years ago skates were fashioned from the bones of elk, ox and reindeer. But although the word comes from the Old North French *escace*, meaning 'stilt', skating probably originated in Scandinavia. By the Middle Ages, it was widely enjoyed as a pastime on the frozen canals of Holland. A Dutch engraving of 1498 depicts the patron saint of skating, St Lydwina of Schiedam, who broke a rib in a fall while skating when she was 16, in 1396. It is the first known illustration of the sport.

If not before, British skaters certainly took to the ice in the Great Frost of 1662. On December 1 of that year the diarist Samuel Pepys recorded seeing people in St James's Park, London, 'sliding with their skeetes, which is a very pretty art'. In 1683, during another Great Frost, Pepys danced on the ice with Charles II's mistress, Nell Gwyn.

In about 1740 the first skating club was founded in Edinburgh and 30 years later the first skating manual, written by Captain Robert Jones, an officer in the Royal Artillery, was published. Knowledge of the sport travelled to North America, where in 1848 E W Bushnell of Philadelphia invented an all-iron skate that clipped onto a boot. The first mechanically refrigerated ice rink, the Glaciarium, opened in London in King's Road, Chelsea, in 1876.

The Dutch developed competitive speed skating in the early 19th century, and it became popular in the Fenlands of Britain where a contest was first held in 1814. Half a century later Jackson Haines, an American dancing master, demonstrated a skating technique that used dance movements, from which figure skating evolved.

TABLE GAMES

Mary, Queen of Scots, who ruled from 1542 to 1567, enjoyed billiards, which was probably known from the 15th century. When billiards was described in 1598 by John Florio, an Anglo-Italian writer, as 'a kind of play with balles upon a table', the cue was flattened into a spoon shape at one end for hitting the ball. It became a tapered rod in the 19th century. In 1875, officers of the Devonshire Regiment invented snooker, named from the slang word for a new cadet at the Royal Military Academy, Woolwich, London.

Archery and darts
Taking aim

The original arrows, wooden sticks with stone heads, were hurled for survival, not sport, about 50,000 years ago in North Africa and by 15,000 BC, flexible sinews and sticks had been combined into bows. For millennia the bow and arrow remained man's most lethal weapon.

To ensure their bowmen's skills were well honed, 13th-century English monarchs kept bows and arrows available for practice. Sporting archery grew as the gun gained military supremacy and in 1537 Henry VIII endorsed the foundation of the Fraternity of St George, which conducted both archery and firearm shooting. In his educational book *Toxophilus, the Scole of Shooting*, dedicated to his monarch in 1545, Roger Ascham, tutor to Princess (later Queen) Elizabeth, recommended it as excellent exercise and recreation.

At competitions Tudor archers aimed at a straw mark 45cm (18in) across from a distance of some 144–216m (160–240yd), and attempted to 'make a length' by shooting all their arrows the same distance – this being the object in battle. In butt shooting, a paper disc was mounted on a butt, a mound of earth, some 90–126m (100–140yd) from the bowman.

In the 18th century the foundations of modern archery were laid by Sir Ashton Lever of Alkrington Hall, Manchester, with the assistance of his secretary Thomas Waring, a skilled bowmaker. Lever founded the Toxophilite Society in 1781, after which archery became widely popular.

By the 17th century French children were playing a game called *les dards* using a square piece of paper marked with a bull's-eye and four-feathered darts. The game became an adult favourite, often played in public houses, from the early 20th century. It spread rapidly after the *News of the World* promoted a darts competition in 1927.

Board games

'It takes too long to play, there's no winning post, no finishing line, the rules are too complicated and the players just keep going round and round.'

REJECTION LETTER FROM PARKER BROTHERS TO CHARLES DARROW ABOUT HIS
MONOPOLY GAME, 1933

Board games have been found in Syria and Israel that date back to about 7000 BC: from such entertainments our modern board games evolved. Chess, a game of skill that mimics battlefield strategies, probably originated in north-west India as *chaturanga*, a two-handed game without dice in which real battles were replayed in miniature using pieces representing the fourfold division of the Indian army: infantry, chariotry, cavalry and elephants.

The first mention of chess, as the game *shatranj*, is found in the Persian *Karnamak-i-Artakshatri-i-Papakan*, a romance written in the early 7th century AD about Ardashir, 3rd-century founder of the Sassanid dynasty. Chess travelled, it seems, from India to Persia in the late 6th or early 7th century. Then, with the Arab conquest of Persia in the 7th century, the game went west to Europe by a variety of routes.

Draught version

In 1996 a board marked with eight by twelve squares was discovered in a grave at Stanway in Essex which dates to the 1st century AD. Each player had twelve identical pieces, and a thirteenth that was different. The game is probably a local version of the Roman game of *latrunculi* – a strategic contest that involved trapping the enemy's pieces. This may be a forerunner of draughts, a game of strategy that seems to have been invented to play on a chessboard in the 12th century in southern France. However, Polish draughts, played on a board of ten by ten squares, appeared in Paris around 1725.

Ludo's predecessor, which came from India, was *pachisi*, or *chaupat*, usually played by four players on a cross-shaped board of cloth with arms 30cm (12in) long. Dice or cowrie shells were thrown to determine each move. Pachisi, often called the 'national game of India', is certainly centuries old, although the earliest evidence of it dates to the 16th

century when the Mughal emperor Akbar favoured play on an outdoor court in his palace at Fatehpur Shikri – using 16 beautiful maidens as his playing pieces. The Popular Game of Parcheesi, widely played by the British under the Raj, was registered in London in 1863.

Gyan chaupar, devised in India as a moral lesson for children, was the forerunner of snakes and ladders. Boards date back to the 18th century, but the game itself is probably much older. The Tibetan form of the game, according to tradition, originated in the 12th century AD. The goal – the last square on the board – was Nirvana, the Hindu state of perfect bliss. Each square was inscribed with such concepts as kindness, arrogance or charity. Ladders led from 'good' squares upwards towards Enlightenment, while snakes led downwards from 'bad' squares.

Streets ahead

Monopoly, named from the Greek for 'owning everything', originated as The Landlord's Game. Patented in the USA in 1903 by the Quaker Elizabeth Magie, it was designed as propaganda against dishonest capitalist landlords. Handmade copies of the game circulated in the 1920s. By 1924, when houses and hotels had been added, its moral message was lost.

In 1933 Charles B Darrow, an unemployed heating engineer from Pennsylvania, redrew the board on the oilcloth covering his kitchen table to depict Atlantic City. For counters he used charms, including a top hat, an iron and an old boot, borrowed from his wife's bracelet. Darrow patented his version. After an initial refusal, based on '52 fundamental playing errors', Parker Brothers finally bought the rights to the game in 1935. An adapted English version based on the streets of London was manufactured in the same year by John Waddington, but the game's true origins were suppressed until exposed by a legal investigation in the 1970s.

Scrabble seems to have been invented in its finished form at the first attempt. In 1938 an American architect, Alfred M Butts, submitted for patent the game Criss-Cross, which he had devised in 1931 while he was unemployed during the Depression. Butts determined the values of individual letters by counting the number of times each was used on the front page of the *New York Times*.

In 1946 handmade Scrabble sets with plywood letter squares, fashioned in his garage by Butts' partner James Bruno, a retired government official, went on sale as Lexico. Bruno was already turning out 200 sets a week by the time the US firm Selchow and Richter bought the rights in 1948 and mass production began in earnest – they sold it as Scrabble from the early 1950s. The British version, which was made by J W Spear and Sons, appeared in 1954.

Jigsaws and conundrums
Puzzles in pieces

An 18th-century London engraver and map-maker, John Spilsbury, devised the first jigsaws to help children to learn geography. Spilsbury made 30 different versions of his 'map dissections', gluing prints onto thin sheets of wood before cutting them into segments. By the end of the 18th century puzzle pictures had become more amusing, but jigsaws remained an expensive rarity until the late 19th century, when a die was invented which could stamp the shapes out of a sheet of card. This, and the development of colour lithography, allowed cheap and cheerful jigsaws to be made.

The term 'jigsaw puzzle' was first coined in the United States in the 20th century from the jigsaw, a vertical saw produced in 1793 that could cut sharp curves.

The *loculus*, a 14-piece puzzle devised in the 3rd century BC by Greek mathematician Archimedes, is the oldest known puzzle made up of several pieces. The difficulty of moving the pieces to form a square is described by the puzzle's alternative name, the *stomachion*, meaning 'the problem that drives one mad'.

In a Chinese woodcut of 1780, two courtesans play with *chi-chiao*, 'seven clever pieces', later the Chinese tangram. Each *tan* represented a celestial body, the Sun, the Moon, Mars, Jupiter, Saturn, Mercury or Venus. The pieces could be arranged into hundreds of forms, from a triangle to a bird in flight.

Solving Rubik's Cube, patented in 1976 by the Hungarian Erno Rubik and designed to help architecture students to think three-dimensionally, was a global craze in the 1980s. In 1984 Larry Nichols, a Massachusetts chemist who had patented a similar puzzle in 1972, won a ruling invalidating the patent. But by then the craze had peaked and Rubik's association with the cube was on its way into the dictionaries.

Card games
Packs and suits

In AD 969, Chinese emperor Mu-tsung was said to have spent New Year's Eve enjoying card-like games. These were played with tokens marked with spots to indicate their values, rather like today's dominoes. Chinese 'money cards', which may also originate from the 10th century, were copied from banknotes. Although the first packs to have suits, they did not have court cards.

The Egyptian Mamelukes were making cards by the 13th century, and the oldest surviving fragment is from one of their sets. A Mameluke set of the 15th century comprised 52 cards in four suits: coins, which were similar to those of Chinese cards, polo sticks, swords and cups. Each suit had three court cards: the *malik* (king), *na'ib malik* (deputy king) and *thani na'ib* (second deputy).

Playing cards like those in the Mameluke set have been discovered in Spain where cards, or *naipes*, were known by 1371. Nine years later cards had also reached such cities as Basle, Paris and Florence. Aristocrats amused themselves with hand-painted cards. Other players used cards printed from wood blocks with colour applied with stencils or fingertips. These cards were the first printed materials to be mass-produced.

Names and symbols for the suits varied from country to country, as did the number of cards in a pack. Today's standard British pack originated from France in 1470. French playing cards were the first to contain a queen rather than the deputy king used elsewhere in Europe. Of the suits, *carreaux*, or 'paving tiles', became our diamonds; *trèfles*, or 'clover leaves' became clubs; and *piques*, or 'pikeheads', became spades, named from the Italian for 'sword', *spada*. The final suit of *coeurs*, or hearts, remained constant.

Games in which 'tricks' are won by the highest card have been played ever since cards arrived in Europe. Two-handed whist was a game of rank first played in the 17th century. It developed from English triumph or trump, played by 1522 and the first game to feature a trump suit outranking the others for the duration of the game. Bridge or biritch, the four-handed game of rank destined to become the world's most popular card game, is derived from whist. The first set of rules was compiled around 1885 by a John Collinson.

Primero was fashionable at the court of Elizabeth I. William Shakespeare certainly knew it, and in one of his history plays has Henry VIII playing it on the night of Elizabeth's birth. Three players bet on their hands, with pairs, three of a kind and three of the same suit sought-after combinations. From primero developed such five-card games as brag and French *poque*, based on betting and bluffing. Old poker, which was played with 20 cards, was being enjoyed in Mississippi by 1829, possibly after poque was introduced by French immigrants.

The father of rummy games is *conquian*, which was played in Mexico before it moved north as coon-can into the USA in the 1850s. Gin rummy was first played in New York in 1909 and canasta was introduced around 1940 at Montevideo, Uruguay. Using two packs, including jokers, canasta became a worldwide craze in the 1950s.

Crosswords
Playing with words

The Victorian parlour game of magic square or double acrostic, in which groups of words were arranged so that the same words read vertically and horizontally, inspired the crossword puzzle. Arthur Wynne, who was born in Liverpool but worked on *The New York World*, remembered his grandfather playing magic square and adapted it to include clues and blanks. In 1913 the newspaper published the world's first crossword. 'Torquemada' compiled the world's first cryptic crossword, which was published in 1925 in Britain's *Saturday Westminster*.

Playground games
Children's favourites

In 1560 the Dutch artist Pieter Bruegel the Elder created *Children's Games*, a complex oil painting in which he depicted children amusing themselves with some 55 different pastimes of the day. Most of these games are still enjoyed by children today, from tree-climbing and blowing soap bubbles to playing hide-and-seek, king of the castle and games of marbles and jacks. A remarkable number of the games were ancient even in Bruegel's time. The children of ancient Greece played versions of pig in the middle (*chytrinda*), tug of war (*dielkustinda*) and hide-and-seek (*apodidraskinda*).

Many of these games have been recorded around the world. Blindman's buff was played in ancient Greece and in Rome, where it was called *chalke muia*, or 'brazen fly'. The English name comes from the Old French *buffe*, a 'buffet' or 'blow', and betrays its once more physical aspects.

Hopscotch was enjoyed in ancient Greece, where it may have been based on the myths surrounding labyrinths and mazes. These were symbols of the perplexities of human life. Hopscotch was later adopted by the Church as an allegory of the soul's hazardous journey from Earth to Heaven.

ANCIENT AND MODERN

✣ Variations on noughts and crosses were known in ancient Egypt, Greece, Rome and China, where the game was played by at least 500 BC. It was introduced into Britain by the Normans in the 11th century, when it was known as 'three men's morris'.

✣ Skipping, jumping over a swinging piece of rope which has handles attached to each end, probably dates back only to the 19th century.

✣ Yo-yos were first mass-produced in the 20th century by the American David Duncan, who in 1929 bought the rights to what was then a popular toy in the Philippines. But such toys were also widespread in Greece and China.

YOU'RE IT!

Chasing games, variously called tig, tag, he or it, are linked to ancient fears of being touched by evil or the devil. The idea that players are safe from being turned into the chaser if they touch iron, which was thought in pagan times to guard against evil spirits, or wood, representing the Crucifix, clearly has religious connotations.

The first conkers were empty snail shells suspended on strings. They were only replaced with horse chestnuts in the 19th century. The name of the game may well stem from 'conqueror', referring to an unbeaten conker.

At the climax of the rhyme 'Oranges and lemons', children chant 'Here comes a chopper to chop off your head'. One theory is that this song harks back to the days of public executions when the condemned person was escorted to execution with the church bells tolling. These words may refer specifically to the executions of Anne Boleyn and Catherine Howard, the wives of King Henry VIII.

Activity toys
Invention and exercise

Roller skates made a dramatic debut at a London masked ball in 1760 when Joseph Merlin, a Belgian instrument-maker, glided into the room on his contraptions playing a violin. Merlin promptly smashed into the opposite wall, breaking both his instrument and a mirror, and seriously injuring himself.

Not until 1823 was the idea of roller skates revived, by Robert John Tyers, a London fruit merchant. Tyers had some success with his Volitos, which had five small wheels arranged in a line, much like Rollerblades, launched in 1980. The traditional four-wheeled design was adapted by American surf-shop owner Bill Richards in 1958 to create the skateboard.

The Frisbee owes its invention to William Russell Frisbie, who in 1871 founded the Frisbie Pie Company in Connecticut. His pies were popular with students at Yale, the local university, who had fun throwing the empty saucer-shaped metal containers to each other. In 1948 Fred Morrison, a Los Angeles building inspector, produced a plastic version. Originally he named it Morrison's Flyin' Saucer, but later changed it to Frisbee, altering the spelling of the creator's name to avoid legal difficulties.

THE KITE, NAMED IN ENGLISH AFTER THE BIRD OF PREY, WAS BEING FLOWN IN CHINA AS EARLY AS 1080 BC, WHEN IT MAY HAVE BEEN USED TO FRIGHTEN AWAY EVIL SPIRITS. ARCHYTAS OF TARENTUM, A GREEK SCIENTIST, BUILT KITES IN THE 5TH CENTURY BC, BUT THE IDEA DID NOT REACH EUROPE UNTIL THE LATE 16TH CENTURY, WHEN THEY WERE BROUGHT FROM THE EAST BY DUTCH TRADERS.

Books for children
Tell me a story

'What is the use of a book,' thought Alice, 'without pictures or conversations?' When Lewis Carroll put these words into the mind of his heroine in *Alice's Adventures in Wonderland*, published in 1865, such thinking was still quite new. Books written specifically for children were rare before the mid 18th century. The earliest example in the English language, an eight-page collection of rhymes entitled *A Booke in Englyssh Metre, of the great Marchante Man called Dives Pragmaticus, very preaty for Children to reade*, was produced by Alexander Lacy in 1563. Almost a century later the Czech educationalist John Comenius introduced the first picture book, *Orbis Sensualium Pictus* (*The Visible World in Pictures*), in 1658, published in Latin and German.

Increasing literacy among the young made specialised children's publishing commercially viable. The publisher John Newbery launched his business in 1744 with the games book *A Little Pretty Pocket Book intended for the Instruction and Amusement of Little Master Tommy and Pretty Miss Polly* and was responsible for many innovations.

In 1751 Newbery introduced the first children's magazine, *The Lilliputian Magazine; or the Young Gentleman and Lady's Golden Library*, a miniature monthly journal containing stories, jokes and songs. In 1765 he published the first novel written specifically for children, *The History of Little Goody Two-shoes*. The anonymous author of this rags-to-riches story is thought to be the novelist and playwright Oliver Goldsmith, an associate of Newbery.

Children's toys
Fun and games

In Mohenjo-Daro in the Indus Valley about 4,000 years ago a child pulled along a model cart and its team of oxen. Other archaelogical discoveries reveal that Egyptian children played with cats with movable tails, miniature snapping crocodiles and figures of bakers kneading dough. To a child from antiquity, a modern toyshop would contain many familiar favourites.

Before the 16th century, when a toy industry developed in Germany, parents and children themselves fashioned toys from whatever materials they had to hand. German toy makers were so successful that most shop-bought toys owned by the children of Europe over the next 300 years were made in their country. Typical toys were wooden animals, often made as sets to furnish farms and Noah's arks.

German craftsmen also applied their skills to metalwork. Miniature soldiers were made in the 13th century – woodcut prints show boys playing with model jousting knights. Most of these early figures were cut-outs from metal sheets. They became rounded only in 1893 when a London firm, Britains, used a hollow-cast moulding process to make model soldiers from the Life Guards cavalry.

Coaches and horse-drawn transport were common toys in the 18th century, and model trains, including some propelled by steam, followed in the 19th century. American companies became the first to produce clockwork train sets in 1855, and electric models in 1884. Pedal cars for children appeared in 1905, less than 20 years after motor cars were first made. Construction sets for aeroplanes were available from 1913, a decade after the Wright brothers' first flight.

BUILDING BLOCKS

In the 19th century, an understanding of the importance of play to a child's development led to the creation of construction toys. Wooden blocks that could be assembled into towns and castles were produced in Germany from 1800, and by 1850 American parents could buy 94-piece miniature log cabins for their children to build.

In 1901 a Liverpool meat importer's clerk, Frank Hornby, patented a system based on interchangeable parts, thin strips of perforated metal held together with nuts and bolts. Marketed at first as 'Mechanics Made Easy', the system became even more popular in 1908 when it was renamed Meccano. One of the most successful construction toys was the Lego bricks system, named from the Danish *leg godt*, 'to play well'. It was devised in 1955 by the Danish carpenter and toy-maker Ole Kirk Christiansen.

Soft toys
Rag dolls and stuffed animals

A stuffed doll found in a Roman child's grave from the 3rd or 4th century BC is probably the earliest surviving rag doll. The person credited with first producing rag dolls on a commercial scale is Margarete Steiff, a wheelchair-bound German seamstress, who began a mail-order business selling toy elephants made of felt, in 1880. The Steiff firm also claimed to have introduced plush bears, a cuddly version of the clockwork bears that were popular in the 19th century, at the Leipzig Fair of 1903.

Steiff bears, which from 1905 featured the famous company trademark of a metal button in one ear, were immediately popular, but the first teddy bears were created in 1902 in the USA. Morris Michtom, a sweetshop owner, and his wife Rose began making stuffed toy bears after seeing a newspaper cartoon of a incident in which Theodore Roosevelt, the American president, spared the life of a bear cub while out hunting. They named the bear 'Teddy' after the president's pet name. The Michtoms were soon running a successful business making toy bears and the Ideal Toy Company remained in the family's control until the 1970s.

MODEL PROGRESS

❋ Talking dolls, who could say 'mama' and 'papa', were introduced in Paris in 1823 by the German musician Johann Maelzel.

❋ William Harbutt, an art teacher in Bath, invented a putty-like modelling material called Plasticine in 1897 after his students complained that clay dried too fast for easy working.

❋ In 1932 Meccano Ltd produced six miniature vehicles as accessories to its Hornby train sets. The cars were so popular that the range was relaunched in 1933 as Dinky cars, from the Scottish slang for 'tiny'.

Dolls and doll's houses
Playing with little people

The earliest known doll-like figurines, including the *ushabti* made by the ancient Egyptians, were not children's playthings but religious talismans which were used as funeral figures. From around 2000 BC these ushabtis represented servants who, in earlier times, would have been buried alongside their dead masters to serve them in the afterlife.

In the late 17th century the habit of dressing dolls in elaborate copies of the day's fashions was established. One early example, a jointed wooden doll made in about 1690, is said to have been owned by the family of James Edward Stuart, pretender to the British throne.

Small girls in ancient Greece are known to have played with jointed dolls made of baked clay. On reaching marriageable age at around 12, a girl would ritually abandon such toys and dedicate them to Artemis, goddess of the Moon and of fertility, a custom later adopted by Roman girls.

Little is known about the dolls that children would have cherished in the Europe of the Middle Ages, but they were certainly commonly used as playthings by the 15th century. German woodcut prints of 1491 depict doll-makers at work fashioning figures out of wood, and some of the dolls even have movable limbs.

MINIATURE MANSIONS

The earliest known doll's house, built in 1558, was commissioned by Duke Albrecht V of Bavaria. Although the house was originally intended as a gift for his daughter, the duke found it so fascinating that he kept it on show in his private art collection. In Britain the oldest surviving doll's house dates from the 1690s and was given to Ann Sharp, daughter of the Archbishop of York, by Princess, later Queen, Anne, who was her godmother.

Dolls were known as 'toy babies' or 'babies' until the 18th century, when the word 'doll', a diminutive of Dorothy, first came into use. However, the first dolls to be fashioned in the image of an infant did not appear until about 1850, when wax models were made in England by Augusta Montanari and her son Richard.

'Wax dolls', with wax heads attached to wooden or cloth bodies, had been introduced in Germany during the 17th century. Mass-production techniques were first used for making dolls in the 19th century, when moulded heads made of papier-mâché or ceramic were turned out in their hundreds.

Dolls moulded from Celluloid, an early form of plastic, appeared in the USA in 1863. Twentieth-century additions include 'teenage' dolls such as Mattel Toys' Barbie. Named after 17-year-old Barbara Hondler, the daughter of the company's founder, she was launched in 1959. Barbie's first outfit was a zebra-striped swimming costume.

CHAPTER 8

SICKNESS AND HEALTH

We think of reconstructive plastic surgery as a marvel of modern medicine, but in the 6th century bc the Hindu surgeon Susruta was using skin grafts to repair patients' damaged ears and noses. His knowledge passed to the Arabs, and then to the Greeks and Romans. Grafting was revived in the 16th century by Gaspare Tagliocozzi, an Italian surgeon who realised that only the patient's own skin tissue could be used. Over the centuries that followed a greater understanding of tissue rejection grew and by the mid 20th century doctors knew enough to be able to attempt the first organ transplants.

The origin of much of today's medicine lies in history: as doctors develop new cures and treatments they rely heavily on experiments and discoveries made in both the recent and distant past. We can see the beginnings of an understanding of how infectious diseases are spread as early as the 1st century bc, when Marcus Terentius Varro suggested that minute particles entering the body caused illnesses. Sophisticated anaesthetics make extraordinary operations possible now, but the ancient Greeks and Romans put patients to sleep before surgery with an extract of mandrake root. And the blood transfusions that save lives today all began in 1667, when Jean-Baptiste Denys gave a transfusion of lamb's blood to a gravely sick boy.

The first diseases
Infections and their origins

Some of the bodies that the ancient Egyptians preserved by mummifying bear witness to such deadly diseases as tuberculosis, smallpox, bilharzia and poliomyelitis. The Egyptians also made note of deaths from the 'plague', a word that was then used to describe a variety of fatal epidemics.

Ever since humans have been on Earth, they have suffered from infectious diseases caused by viruses, fungi, bacteria, protozoa and parasitic worms. Some of the first human diseases were caught from the other primates with whom our early ancestors shared their tropical habitats. Malaria and yellow

fever are two of our oldest killers and were spread by flying insects. They became infected by feeding on primate blood, then presumably passed on the infection to humans by biting them.

About 250,000 years ago humans began moving to temperate regions. Here there were no other primates from which diseases could be caught, but as humans huddled in caves for warmth and shelter they became hosts to parasites such as bed bugs, which had previously only infected cave-dwelling bats. Those infections able to spread themselves by physical contact or in the spray of coughs and sneezes now had a perfect opportunity to proliferate. Such diseases became even more successful from about 12,000 years ago, when, with the beginning of agriculture, more humans began living at close quarters.

Crowded living conditions also proved the ideal breeding grounds for tuberculosis and typhoid. Rats and their fleas became vectors of bubonic plague, a disease which was first described by the Roman physician Rufus of Ephesus during the 2nd century AD.

Understanding illness
Cause and effect

WHILE INVESTIGATING WHY BEER AND WINE SPOIL, THE FRENCH CHEMIST LOUIS PASTEUR DISCOVERED THAT AIRBORNE ORGANISMS ARE RESPONSIBLE AND THAT THEY CAN BE DESTROYED BY HEAT. IN 1885 HE PRODUCED A RABIES VACCINE BY HEATING BACTERIA. THUS WEAKENED, THE BACTERIA COULD BE INJECTED INTO PATIENTS TO MAKE THEM IMMUNE.

Until the 19th century, most illnesses were attributed to divine displeasure or to low-quality air – malaria was named from the Italian for 'bad air'. However, some diseases, such as leprosy were known to be 'catching'. It was understood that leprosy was contagious in Old Testament times, when all lepers were declared 'unclean' and forced to live in isolated colonies.

In the 1st century BC the Roman encyclopedist Marcus Terentius Varro speculated that disease might be caused by minute particles entering the body, and in the 6th century AD the Hindu doctor Susruta suggested that malaria might be spread by mosquitoes. But the realisation that people or things could transmit 'plagues' was not accepted until the Middle Ages. From about 1380, after the plague or Black Death had exterminated nearly a quarter of Europe's population, ships carrying infection were refused entry to Venice. At Ragusa on the Adriatic, in the days of the Black Death, immigrants and traders had to remain outside the city for 40 days to prove they were not infected. This was known as *quarantinza*, from the Italian *quaranta*, or 'forty'.

Epidemics, suggested the Italian physician Giraolamo Fracastoro in 1546, were caused by some kind of 'seeds' that were wafted through the air or carried in water. Much later the microscope would help to confirm this principle, but as early as 1683 the instrument's Dutch inventor Anton van Leeuwenhoek probably viewed bacteria from his teeth by means of its powerful lens. In 1850 Casimir Davaine showed that anthrax could pass in the blood from infected to uninfected sheep and cattle, and detected rod-shaped anthrax bacilli in newly infected animals.

The British anaesthetist John Snow showed that cholera was spread in water, and in 1854 stopped a cholera outbreak in London by removing the pump handle of a polluted public well.

Even the most powerful microscopes could not detect viruses, the minute agents responsible for diseases such as polio. In 1892 the Russian bacteriologist Dmitry Ivanovski discovered that the agent of tobacco mosaic disease, which affects tobacco leaves, was not filtered out by a fine mesh that trapped bacteria. In the late 1930s, when the electron microscope was invented, viruses were finally observed directly.

Preventing infection
Inoculation and vaccination

In a practice that may originally have come from India, the Chinese of the 11th century AD put the scabs from pustules of smallpox victims into the nostrils of non-sufferers to give them a mild form of the disease and so prevent a severe attack. Although fatalities numbered 1 in 50 at best, this 'elective infection' quickly became widespread after being introduced into Britain in 1720.

By the 18th century it was well known in country districts that milkmaids who caught cowpox did not succumb to smallpox. Using a cobbler's needle, a farmer named Benjamin Jesty 'injected' his wife and family with pus from the udder of a cow infected with cowpox in 1774.

True vaccination was first achieved by the British physician and naturalist Edward Jenner. In 1796 Jenner encountered Sarah Nelmes, a dairymaid with fresh cowpox lesions on her finger. A few days later he inoculated eight-year-old James Phipps with pus from the dairymaid's pustules. The boy developed cowpox symptoms.

After a few weeks Jenner risked injecting Phipps again, this time with part of a human smallpox scab. To Jenner's relief and delight the child remained well. In 1798, after further successes, Jenner published his results to public acclaim. Further vaccines – named later from *Vaccinia*, after the cowpox virus – for other diseases, were to follow.

The story of vitamins
Vital ingredients

In 1912 Casimir Funk, a Polish-born American biochemist, coined the word 'vitamine' for a group of substances he thought to be 'vital' to life. Funk also believed these vital 'amines' were chemicals derived from ammonia. The first part of Funk's theory was proved correct, the second partially so, which is why 'vitamin' lost its final 'e'.

In his journeys of the 1770s, which included a landing in Tahiti, Captain Cook preserved the health of his men by insisting that they ate orange extract, sauerkraut and cress. Citrus fruits were later discovered to contain large amounts of vitamin C.

Scurvy was an occupational hazard for sailors, although from the 1500s Dutch seamen benefited from including citrus fruits in their diet on lengthy voyages. Even if it did not kill him, scurvy made a sailor's teeth fall out, weakened his bones and prevented his wounds from healing.

By the time the British naval surgeon James Lind wrote his *Treatise of the Scurvy* in 1753, more British sailors succumbed to the disease than died in action. When Lind's advice that British sailors' diet should contain fresh citrus juice or fruit was adopted by the Royal Navy in 1796, scurvy vanished. What Lind did not know was that scurvy is caused by lack of ascorbic acid, or vitamin C, which citrus fruits contain.

In 1901 Gerrit Grijns showed that polishing rice removed an essential nutrient. Eating brown rice with some of the husk left on could prevent beriberi. The disease, known in the East for millennia, could cause fatal paralysis, emaciation and anaemia.

Mental illness
Diseases of the mind

Until the 19th century, mental illnesses were usually attributed to the power of evil spirits or the displeasure of the gods. Sufferers were routinely exorcised, banished, punished or locked up in prisons or asylums. But Greek physicians recognised a disease they named *apoplexia*, described by the Roman Caelius Aurelianus as a 'sudden collapse, as if from a deadly blow ... in general without fever, and it deprives the body of all sensation.'

In about 400 BC the Greek physician Hippocrates linked apoplexy with black bile or melancholy, one of the four humours, and noticed that it was prevalent in rainy weather. He and his successors prescribed a regime of calm and 'occupation', plus hellebore and other purgative drugs. But such sympathetic attitudes had disappeared in western Europe by the Middle Ages, when witchcraft and demonic possession were again deemed the only reasonable explanations for insanity.

From the 8th century Muslim Arabs set up asylums for 'retreat and security'. The first hospital for the mentally ill had been founded in Spain by 1409, but by the 17th century most of Europe's mental patients were chained to the walls of dark 'dungeons'. The French physician Philippe Pinel, director of the Bicêtre asylum for men from 1793, removed patients' shackles and talked to them about their problems.

The idea of 'talking cures' became, in the late 19th and early 20th century, central to the work of Sigmund Freud, the Viennese neurologist. In his *Studies on Hysteria* of 1893–95, regarded as the first written account of psychoanalysis, Freud described how nervous diseases, typified by symptoms such as paralysis and 'fits', could be linked to such drives and emotions as sexual desire and guilt buried deep in the unconscious mind.

MANY OF FREUD'S PATIENTS WERE ANALYSED UNDER HYPNOSIS, A TREATMENT PRACTISED BY THE ANCIENT EGYPTIANS TO CHANGE STATES OF MIND. THE TERM 'HYPNOTISM', FROM THE GREEK *HUPNOS*, 'SLEEP', WAS COINED IN 1843 BY THE SCOTTISH SURGEON JAMES BRAID TO DESCRIBE TRANCES INDUCED BY THE AUSTRIAN PHYSICIAN FRANZ ANTON MESMER.

Hormone treatment
Understanding diabetes

During the 16th century BC, the Egyptians made medicines 'to drive away too much urine'. One such prescription was a liquid strained, after four days, from a mixture including cakes, wheat grains, green lead, soil and water. Aretaeus of Cappadocia, a physician of the 2nd century AD, was the first to describe diabetes accurately. Because of sufferers' frequent urination, he named it after the Greek word for 'siphon'. Aretaeus also recognised the thirst 'as if scorched by fire' and 'dreadful emaciation' typical of diabetes.

For another 2,000 years diabetics continued to waste away because their pancreas glands were not releasing insulin, the hormone essential to the control of energy-supplying glucose from their blood. Thomas Willis, physician to Charles II, made a step towards identifying the cause in 1674 when he described the urine of diabetics as 'wonderfully sweet'. Thereafter the adjective *mellitus*, from the Latin 'honey-sweet', was added to the name.

In 1869, German student Paul Langerhans discovered the existence of thousands of minute tissue clusters in the pancreas. These were later named the islets of Langerhans in his honour. In 1889 two Germans, Oskar Minkowski and Joseph von Mering, discovered that the islets had some connection with diabetes, and from 1909 the term 'insulin', from the Latin *insula* ('island'), was being used for the islets' secretions.

Attempts were made to extract insulin from the pancreas, but it was the pioneering work of Charles Best and Frederick Banting in Canada that made insulin treatment possible. Banting and Best isolated insulin in 1921, and in January 1922, having injected each other to ensure the safety of their procedure, they administered it to their first patient, 14-year-old Leonard Thompson. The boy, who had been close to death, made a total recovery. Thereafter he took regular doses of insulin.

WHAT AILS YOU

✻ Arthritis has been detected in Neanderthal skeletons 50,000 years old. The ancient Egyptians called it 'hardening in the limbs'.

✻ Epilepsy is named from a Greek word meaning 'a taking hold of'. Sufferers were thought to be seized by mysterious powers.

✻ When sailors' daily supplies of lemon juice were replaced by lime juice in the mid 1800s, American sailors called their British counterparts 'limeys'.

✻ Vitamins with similar functions were first grouped by code letters in the early 20th century.

✻ In 1943 the Dutch physician Willem J Kolff treated a patient with the first artificial kidney, a cellophane filter immersed in a water bath that was linked to the patient's bloodstream.

Guardians of health

'I swear by Apollo the physician, and Asclepius, ... I will follow
that system of regimen which, according to my ability and
judgment, I consider for the benefit of my patients, and
abstain from whatever is deleterious and mischievous.'

FROM THE HIPPOCRATIC OATH, 4TH CENTURY BC

In early societies shamans were the guardians of a community's health.
Healer priests, who combined medicine with astrology, practised in
Sumeria and Babylonia, but according to Greek historian Herodotus,
writing in the 5th century BC of an era about 2,500 years earlier than his own,
Babylonia had no specialist physicians. He described how the sick were taken
into the marketplace and passers-by asked, on the basis of their
experience, to give advice on treatment. 'No one', he said, 'is
allowed to pass the sick man in silence without
asking him what his ailment is.'

The Code of Hammurabi, a set of
Babylonian laws carved on a stone pillar
in the 18th century BC, encompassed
recommended legal practice for doctors.
Penalties for malpractice were severe. 'If the
doctor, in opening an abscess, shall kill the
patient his hands shall be cut off,' it decreed.

The Ebers papyrus, written in Egypt
in about 1550 BC, describes the medical
practices known at the time. Imhotep, the
work's central character, is the first physician
in historical record. Chief adviser in the Egyptian
court of Zoser, who reigned from 2630 to 2611 BC, Imhotep was
so revered for his healing powers that he was worshipped as a
god. The papyrus also reveals that specialists were common in
ancient Egypt. Although he also took on cases of eye and belly
disease, Iry, a court physician of about 2500 BC, was given the title
'Keeper of the King's Rectum'.

Holy healing

From about 1200 BC the Greeks worshipped Asclepius, the god of
medicine who was identified with Imhotep; temples were built in
his honour throughout Greece. In life, Asclepius was thought to
have been a physician who made some miraculous cures. As Greek
civilisation advanced, philosophers started searching for rational
explanations of illness. In about 460 BC Hippocrates was born in the
Greek island of Cos. Known as the 'father of medicine', Hippocrates

taught and practised in Cos and other parts of the Mediterranean area. He proposed that illness was caused by malfunction of the body rather than the malign effects of supernatural interference.

Medicine advanced rapidly after Hippocrates' time, and a medical school was established at Alexandria in 300 BC. The Greek physician Galen, whose teachings influenced medical practice for more than 1,500 years, gained his first experience of medicine in around AD 140, when, as a boy, he visited the medical school in his home town of Pergamum (now Bergama in Turkey) to study how wounded gladiators responded to treatment.

Like Galen, Greek and Roman physicians received no formal education, and 'apprentices' learned their skills on the job. Before the 3rd century AD, when under the emperor Severus formal training and bedside lessons were introduced, anyone could set up in practice as a physician.

After the fall of Rome, European medicine was kept alive by the Church, with monks gaining medical knowledge in new hospitals such as London's St Bartholomew's, founded in 1123. The first medical school was set up in Salerno, Italy, between the 9th and 11th centuries. Until the 1600s the *Canon of Medicine*, by the Persian philosopher and physician Avicenna, remained the basis of medical teaching in Western hospitals and universities. Written in about 1010 it drew on a 9th-century translation of Galen's works.

Women's work

In ancient Greece, women like Antiochis, whose skills were honoured by her home town of Tlos, practised medicine and it was one of the few male occupations open to Roman women. In China the first record of a female doctor dates to about 160 BC. Throughout medieval Europe some women 'leeches' practised as healers, including Euphemia, the 13th-century abbess of Wherwell in Hampshire. In Italy, Germany and Korea women were allowed medical training from the 14th century, but elsewhere the total exclusion of women persisted.

The modern pioneers faced huge public opposition, but in 1849 Elizabeth Blackwell became the first woman to qualify as a doctor in the USA. Elizabeth Garrett Anderson, inspired by a meeting with Blackwell, passed the exams of the British Society of Apothecaries in 1865.

Measuring pulse and blood pressure
The beat of life

The pulse's throb has been felt since ancient times to assess people's health, and in about AD 1100 the Greek doctor Archimanthaeus wrote in *The Coming of a Physician to his Patient*: 'The fingers should be kept on the pulse at least until the hundredth beat in order to judge of its kind and character …'.

In the 17th-century, Italian professor of medicine, Santorio Santorio, timed the pulse against a 'pulsilogium', an instrument incorporating a pendulum. But only after 1707, when John Floyer of England made use of his 'pulse watch', were rates regularly or accurately timed.

To measure blood pressure, the Italian doctor Scipione Riva-Rocci wound an inflatable bag around the arm of his patient to constrict its main artery. He then used a column of mercury to measure how much pressure in the bag would completely stop the blood flow – as detected by feeling the patient's pulse at the wrist. The device was named a sphygmomanometer, from the Greek word *sphugmos*, meaning 'pulsation'.

Riva-Rocci's invention dates back to 1896. However, it had been preceded 20 years earlier by an instrument combining an aneroid barometer with a bulb pressed against the wrist to constrict circulation. The refinement of using a stethoscope to listen to the sounds in the artery as the blood overcame the pressure in the cuff was added in 1905 by Russian physician Nikolai Korotkoff.

FIRST AID

To bind wounds, the Egyptians used linen bandages spread with honey and myrrh (a mild antiseptic). By 2500 bc they were also using palm tree fibres as splints. Modern sticking plaster was being made in the USA by Robert Shoemaker by 1838. Band-Aid was produced in the 1920s by the US company Johnson & Johnson. The idea came from an employee, Earle Dickson, who had laid sterilised gauze inside a roll of sticking plaster to create ready-prepared bandages for his accident-prone wife.

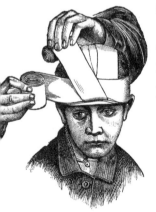

The stethoscope
Listening to the body

By tapping a patient's chest and listening to the sounds produced, Leopold Auenbrugger, an Austrian physician of the 18th century, realised he could gain vital clues about the heart and lungs. Auenbrugger's technique was popularised by the Frenchman Jean Nicolas Corvisart after 1801.

The only other way of detecting internal sounds was to place an ear to the body. In 1816, however, the French physician Theophile René Hyacinthe Laënnec invented the first listening aid – a paper tube he called the stethoscope – in order, reputedly, to avoid having to put his ear on a woman's chest. Laënnec later used a wooden tube 30cm (1ft) long as a listening device. He wrote a book about his work in 1819, offering purchasers free stethoscopes. The flexible, twin earpiece design of the 1850s used Laënnec's principles.

X-rays and endoscopes
Looking inside

Before the end of the 19th century a doctor could detect changes inside the body only by cutting it open. But diagnosis was revolutionised by the German scientist Wilhelm Roentgen. While passing electrical discharges through vacuum tubes on November 8, 1895, he discovered that a nearby piece of paper that was painted with the substance barium platinocyanide glowed with a bright fluorescence.

Roentgen then found that invisible rays from the tubes, which he called X-rays because he did not know what they were, blackened photographic plates. He used these plates to show that when the rays passed through the body they were blocked to different extents, most of all by bone. Roentgen created the first permanent X-ray image, of the bones inside his wife's hand, on December 22, 1895.

Within weeks an astonished world was shown X-rays. One London newspaper reported how they passed through wood, wool and flesh to create pictures of Frau Roentgen's hand bones and ring. 'They are', it said, 'ghastly enough in appearance, but from a scientific point of view they open up a wide field for speculation.' X-ray diagnosis of bone fractures became widespread, and the diagnosis of diseases of soft tissues, such as the lungs, was also revolutionised by Roentgen's work.

A tube lit by a candle formed the first endoscope, named from the Greek *endon*, 'within'. It enabled 18th-century physicians to look into the rectum to detect disease. In 1868 Adolf Kussmaul, the German physician, passed a rigid tube into the stomach of a sword swallower but he was unable to see anything because the tube was unlit.

The first endoscope fitted with a lens for viewing the stomach was made by the Polish surgeon Joseph von Mikulicz in 1881. It was lit with a modified light bulb. However, the endoscope's potential was only truly realised after 1955 when the Indian physicist Narinder S Kapany, working in London, used bundles of minutely thin glass fibres to carry light down a tube into the body and send back clear fibre-optic images to the viewer.

Heart and brain monitors
Electricity from within

That electric signals were involved in muscle contraction was discovered by the 18th-century Italian physiologist Luigi Galvani. Ordinary electric currents had been measured with galvanometers since the 1820s, but in 1903 the Dutch physiologist Willem Einthoven devised a galvanometer sensitive enough to pick up on the body surface electric signals from a beating heart. These were translated into movements of a needle, which traced an electrocardiogram, or ECG. Einthoven then worked out how heart defects were shown in the traces.

In 1924 Hans Berger, a German psychiatrist, discovered how to detect on the skin of the scalp electrical signals made by the brain. His first 'brainwave' or electroencephalograph (EEG) trace came from his son.

ULTRASOUND TECHNOLOGY WAS FIRST USED AS A DIAGNOSTIC TOOL BY PROFESSOR IAN DONALD IN THE GLASGOW ROYAL MATERNITY HOSPITAL IN 1956. ITS FIRST APPLICATION WAS TO ESTIMATE THE SIZE OF ABNORMAL GROWTHS IN THE OVARIES, BUT WAS SOON BEING USED TO VIEW AND MEASURE UNBORN BABIES IN THE WOMB AND MONITOR FOETAL DEVELOPMENT.

New parts for old

'Only a few months ago I lay in hospital, a dying man with a stricken heart. Then came the miracle. I was given a new lease of life. Today, I am the second man since the Creation to live with the heart of a dead man beating in his breast.'

PHILIP BLAIBERG, SECOND HUMAN HEART TRANSPLANT PATIENT, 1968

The Persian soldier Hegesistratus was the first person known to have worn an artificial limb. In the 5th century BC, Herodotus, the Greek historian, described how Hegesistratus cut off his foot to escape imprisonment by the Spartan enemy. Equipped with a wooden replacement he then fought against them. To replace a hand severed in battle the Roman soldier Sergius Silus possessed an 'iron hand', so Pliny recorded in his *Natural History* of the 1st century AD. And a skeleton excavated in central Italy of around 300 BC had a wooden leg covered with realistically modelled bronze sheeting between knee and ankle.

Simple artificial limbs, such as peg legs or false feet or hands with or without 'hooks', were worn largely for cosmetic effect. They provided some support, but no movement, and changed little for centuries until a 16th-century French military surgeon, Ambroise Paré, started to make hands fitted with holders for quill pens and even with fingers made individually mobile by a series of minute levers. Paré also created ingenious artificial legs of wood or metal. Their pivoting knee joints and sprung feet were designed for wounded soldiers. Some of Paré's elbow joints moved with the help of ratchets.

Miracle materials

Replacement surgery, initially used for damaged and arthritic hips, was pioneered in 1905 by the Chicago surgeon J B Murphy. Progress was slow until the early 1930s when the potential of vitallium, a noncorrosive metal, for oral rebuilding in dentistry was discovered. Knowledge of vitallium's possibilities reached two US surgeons named Venable and Stuck who in 1932 employed it for joint repair.

In 1979 a patient with burns over more than half her skin was the first to be treated with Silastic, an artificial skin made from treated shark's cartilage and cowhide developed in Massachusetts by the surgeon John Burke and Ioannis Yanna, a chemist. The patient's own skin grew into and over the Silastic matrix, which was eventually absorbed by her body. Artificial ligaments made of Goretex, better known as a sportswear fabric, were approved for use in the USA in 1988. Although rarely used as total replacements these are invaluable aids to recovery.

The English surgeon W H Walshe suggested in 1862 that if a heart stopped beating it could be 'shocked' back into action with electricity. The artificial pacemaker, which gives a lifesaving boost to an ailing heartbeat, was named by A S Hyman, an American cardiologist who in 1932 devised a bulky apparatus weighing 7.2kg (16lb) to deliver regular electrical stimulations to the heart. By 1960 a team from New York had inserted the first pacemaker, which was charged from a small unit attached to the outside of the chest.

During the 1950s the first artificial hearts were built in laboratories, but it was not until 1969 that one was implanted, by Denton Cooley of Texas. This air-powered plastic device was a temporary replacement to keep patients alive until hearts became available for transplant. In 1982 an American dentist named Barney Clark survived for 112 days after receiving the first artificial heart intended as a permanent fixture.

By the 6th century BC the Hindu surgeon Susruta was practising plastic surgery in India. He used skin from other parts of the face to create new ear lobes and noses. Susruta's knowledge passed to the Arabs and to the Greeks and Romans. Skin grafting was revived in the 16th century by the Italian surgeon Gaspare Tagliocozzi, who attempted to 'restore the appearances of patients who had lost their noses'. Although he had no knowledge of tissue rejection, Tagliocozzi realised that only tissues from the same individual would 'take'.

Organ transplants

In 1823 a G Bunger of Germany rebuilt part of a nose with skin from a patient's own thigh. Experiments with bone and skin grafts advanced in the following century, and the first transplant of a major body organ took place in 1950 when a patient was given a new kidney by the American surgeon Richard H Lawler. Four years later an American team led by Joseph Murray transplanted a kidney from one identical twin to another. Such successes highlighted the significance of rejection, identified in the 1950s by the US physician Emile Holman as a reaction of the body to 'alien' tissues. The identical twins' tissues were accepted as 'self' because they shared the same genetic make-up.

Throughout the 1950s and 1960s the essentials of rejection were gradually worked out. Now the way was clear for the first liver and lung transplants, which took place in 1963. In December 3, 1967, the South African surgeon Christiaan Barnard gave a new heart to 55-year-old Louis Washkansky at the Groote Schuur Hospital in Cape Town. Washkansky lived for 18 days.

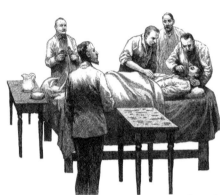

Early surgery
Ancient operations

Using sharpened stones, surgeons of some 10,000 years ago drilled holes in their patients' skulls, probably to release the spirits thought to be causing mental illness. Ancient skulls from around the world show similar signs of this trepannation operation, which Greek physicians also used to treat head injuries.

Basic operations, including setting fractures and dislocations, removing thorns and probably stitching wounds, date back to ancient Egypt and Mesopotamia. But because breaching the body's defences frequently led to fatal infections, surgery was not only minimised but confined to the extremities of the body.

From the time Hippocrates practised in the 5th century BC the Greeks knew how to stitch torn muscles, tie off burst blood vessels and remove bladder stones. Techniques were improved, and surgery's prestige enhanced, by the Greek physician Galen around AD 150.

Following the fall of Rome, surgery became the work of pedlars, magicians and rat-catchers. And even in the mid 1700s, barber surgeons were still performing routine operations. An exception was Ambroise Paré, a 16th-century French surgeon. In Paré's time, the use of gunpowder was increasing. Missiles fired from guns tore into human flesh and pushed what was thought to be 'gunpowder poison' (but was in fact metal and other materials) deep into wounds. Such injuries were routinely sealed or cauterised with hot oil, but in 1536, during the siege of Turin, oil supplies ran out. Instead, Paré used a mixture of egg yolk, rose oil and turpentine, and discovered that with this gentler treatment wounds were less painful and inflamed.

Using anaesthetics
Freedom from pain

ANCIENT GREEK AND ROMAN SURGEONS STITCHED UP PATIENTS' WOUNDS WITH ANIMAL GUT. IN ANCIENT INDIA, SURGEONS STITCHED INTESTINES USING BENGALI ANTS. PLACED SIDE BY SIDE ALONG A WOUND, THE CREATURES CLAMPED IT SHUT WITH THEIR JAWS. THE SURGEON THEN SEVERED THE ANTS' BODIES, LEAVING THEIR HEADS AND JAWS IN PLACE, AND STITCHED UP THE OUTER TISSUES. AS THE INTERNAL WOUND HEALED, THE ANTS' HEADS SIMPLY DISSOLVED.

To deaden the excruciating pain of surgery the Chinese gave patients a mixture of wine and herbal drugs, probably narcotics derived from the mandrake plant. But it was also thought honourable to ignore pain, and operations were regularly performed without anaesthesia. The Greeks and Romans also used mandrake extracts. According to the Roman writer Pliny: 'When the mandrake is used as a sleeping draught the quantity administered should be proportional to the strength of the patient.'

Opium was used to quell pain in ancient times and long beyond, as was alcohol, although it often made patients fighting drunk. In 1844 an American dentist named Horace Wells tried using nitrous oxide for pain relief, discussing its problems with a dentist colleague, William Morton. Deciding that ether would be more effective, Morton used it in 1846 to anaesthetise a patient undergoing surgery to remove a neck tumour. Sir James Young Simpson, a British obstetrician, used ether in 1847 before discovering that chloroform was more effective in relieving the pain of childbirth.

Preventing infection
Killing germs

Following surgery, patients' wounds commonly festered and swelled up, filled with pus, now known to be caused by the immune system attacking infective bacteria. For generations this was regarded as an encouraging sign that healing was in progress, and in the Middle Ages pus was commonly hailed as 'laudable'.

The vital link between disease and pus was not realised until the surgeon Joseph Lister carried out the first antiseptic operation, which he did in 1865 on an 11-year-old boy. By spraying the operating theatre with carbolic acid (a chemical he knew as a sewage disinfectant), and dabbing it on the wound to make an 'artificial scab', Lister killed germs – carried from place to place by direct contact and in the air. Suffering no ill effects, the boy healed quickly.

Antisepsis was effective but messy. Asepsis, in which germs are kept away from the wound during and after surgery, rather than being killed after infection has begun, began to supersede it after the French microbiologist Louis Pasteur suggested in 1874 that surgical instruments could be sterilised by placing them in boiling water.

BLOOD MATCH

Aboriginal Australians may have practised blood transfusion for thousands of years, but not until 1667 did Jean-Baptiste Denys, surgeon to Louis XIV, make the first documented transfusion. He transfused lamb's blood into a young boy weakened by the blood letting of other doctors. In 1825 James Blundell made a successful human-to-human transfusion in London, but the importance of blood compatibility was first realised by Karl Landsteiner, an Austrian pathologist. He identified groups A, O, B and AB in 1900, and from 1908 blood grouping was routinely ascertained before transfusions took place.

Tools and techniques
The art of surgery

Crude flints, or knives made from the hard volcanic glass obsidian, were man's first surgical implements. About 4,500 years ago copper blades were used in Egypt for operations such as male circumcision, while wounds were stitched with copper needles. Sumerian surgeons employed similar tools, together with saws for cutting bone.

The only full set of surgical apparatus surviving from these times is a set of tools found in a Minoan tomb on the Greek island of Crete. The copper instruments in this 3,500-year-old kit include such familiar medical items as forceps, drills and scalpels, as well as a dilator for internal examinations.

Roman surgeons performed their operations with cutting blades, or *scalpelli*, crafted from the best Austrian steel, which remained unmatched until the 18th century. They also favoured catheters – S-shaped tubes used during the 3rd century BC by the Greek doctor and anatomist Erasistratus for treating genitourinary blockages.

Nature's medicine chest

'The Lord hath created medicines out of the earth;
and he that is wise will not abhor them.'

ECCLESIASTICUS, CH. 38, V. 4

The bark of the Pacific yew tree, *Taxus brevifolia*, was discovered in 1992 to contain taxol, a substance that kills cancer cells. Like this new drug, our first medicines came from plants. When, about 60,000 years ago, a Neanderthal man died in Shanidar (now in Iraq) his body was covered with flowers, many of which are still used in local medicines.

The Chinese were the first to record their plant remedies, from about 3000 BC. The herbal *Pen Tsao*, probably compiled during the reign of emperor Fo Hi or his son Shen Nung, dates to this period. It describes more than 260 herbal medicines, among them *Ephedra*, whose active ingredient, ephedrine, is still valued for treating asthma.

Ancient Egyptians and Sumerians made use of health-giving plants, and an Egyptian papyrus of 1500 BC lists hundreds of medicinal herbs. Garlic was chewed by the slave workers engaged in pyramid-building to help to ward off infections and fevers.

Plant power

The Greek physician Hippocrates described some 400 herbal remedies, while much knowledge was spread across Europe by the Romans. As they settled in conquered lands, the Romans planted their favourite medicinal herbs: mint to stimulate the appetite, liquorice to aid digestion and mustard for chest complaints.

Following the collapse of the Roman Empire, Arab physicians became the guardians of herbal knowledge. A treatise written in the early 11th century by Avicenna, a Persian doctor who had learnt how to distil essential oils from plants, became the standard work of Arab medicine. In the West, medieval monks planted 'physick' gardens with herbs that had been introduced by the Romans. But such practices remained largely unknown until herbals were published from the early 1500s. The herbal published in 1654 by the English apothecary Nicholas Culpeper remains in print today.

By 300 BC the Egyptians were using foxgloves to make heart medicine. The Romans employed the foxglove as a diuretic, but the plant's healing potential was not fully tested until the 18th century when William Withering, an English doctor, began searching for a remedy for dropsy, a type of water retention caused by heart disease. In about 1775 he encountered a cure, concocted by an old woman, that made patients very sick. Having identified the foxglove as the cause of the vomiting, Withering went on to investigate its properties. Ten years later, in

An Account of the Foxglove and some of its Medical Uses, he noted the herb's 'power over the motion of the heart'. Withering did not then connect dropsy with heart failure, but discovered that a preparation of foxglove leaves could be used to strengthen the heartbeat.

In ancient Peru *Cinchona succirubra* was known as the 'fever tree'. The bark, now known to contain quinine, was used to cure malaria, a disease endemic throughout Europe and Asia by the 17th century. A missionary reported on the bark's powers in 1633, but these remained generally unknown in Europe until a Spanish Jesuit missionary took this 'Jesuit's bark' to Rome in the 1630s.

Willow tree bark was probably chewed by country folk to calm fevers long before English pharmacist Edward Stone used a powdered form to treat ague. More than a century later, in 1876, Scottish physician Thomas Maclagan gave rheumatic fever patients salicylic acid made from the bark. His patients did well, but the drug caused gastric problems. After Felix Hoffman of Bayer in Germany synthesised pure acetylsalicylic acid in 1897 such side effects were eliminated. From 1899 Bayer marketed the drug as aspirin powder.

Smoking cures

When Christopher Columbus and his men arrived in Cuba in 1492 they were astonished to find local people smoking cigars – probably named from *sik'ar*, a Mayan Indian word meaning 'smoking' – made from rolled corn husks filled with shredded tobacco. 'The effect', reported one of the explorers who sampled it, 'is a certain drowsiness of the whole body accompanied by a certain species of intoxication …'.

Tobacco was brought to Europe by the Spanish, who landed in Mexico in 1518. They coined the name *tabaco*, from the Arabic *tabaq*, meaning 'euphoria-inducing herb'. The word was not recorded in English until John Hawkins visited Florida in 1565. After Hawkins brought tobacco home with him the smoking habit caught on. Soon tobacco was being recommended as an aphrodisiac and a cure for diseases from 'superfluous phlegm' to lockjaw.

By the 4th millennium BC the potent effects of opium had been widely discovered. The Sumerians wrote enthusiastically of the 'joy plant' and probably introduced this poppy to the Egyptians. Cities such as Thebes were famed for their poppy fields by the 15th century BC, when Egyptians were being entombed with opium supplies for comfort in the afterlife.

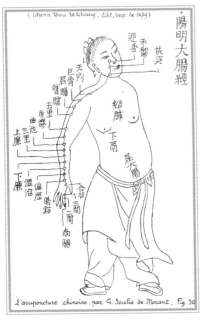

(Tcherin Toiou ta tchzeng . Edit. Imp. de 1679)

陽明大腸經

L'acuponcture chinoise, par G. Soulie de Morant. Fig. 30

The art of acupuncture
Healing with needles

When, according to ancient Chinese philosophy, the body is in harmony, the contradictory forces of yin and yang are so well balanced that they allow *chi*, the life force, to run through it unhindered. Chi is believed to travel along twelve channels or meridians linked to internal organs. At certain places on the channels an imbalance of chi can be regulated and in acupuncture, which may have been devised to release evil spirits, needles are inserted into these points.

Acupuncture was probably devised around 5,000 years ago and performed with stone needles. Techniques were imparted orally until Huang Ti, the 'Yellow Emperor' of about 2600 BC, compiled his *Nei Ching, Su Wen* (*Classic of Internal Medicine*), which included a guide to acupuncture. Around this time the first therapeutic success was recorded after a comatose patient's life was saved. *The Illustrated Manual Explaining Acupuncture and Moxibustion with the Aid of the Bronze Figure and its Acu-points*, written in AD 1027 by Wang Wei-I, was responsible for refining acupuncture. It was accompanied by a pair of life-size bronze figures on which the acupuncture points were accurately marked. Moxibustion involves placing cones of dried mugwort leaves on the skin and setting them alight. Wang Wei-I's book was copied onto a pair of huge stone slabs. Copies, taken from it like brass rubbings, helped to ensure the introduction of acupuncture into Europe in 1683 by Willem tem Rhijne, a Dutch doctor who also coined the name acupuncture.

Homeopathic remedies
Curing like with like

BRONZE NEEDLES FASHIONED DURING THE 8TH CENTURY BC HAVE BEEN UNEARTHED. BUT THE OLDEST SET UNDOUBTEDLY USED FOR ACUPUNCTURE, WITH FIVE SILVER NEEDLES AND FOUR GOLD, MAKING THE STANDARD COMPLEMENT OF NINE, WAS DISCOVERED IN THE TOMB OF PRINCE LIU SHENG, WHO DIED IN 113 BC IN HUBEI PROVINCE.

It was his disillusion with orthodox medicine that led Samuel Christian Friedrich Hahnemann, a medical student at Leipzig University in the 1770s, to search for an alternative means of treating the sick. Hahnemann devised his system of homeopathy, named from the Greek words *homos*, or 'same', and *pathos*, or 'suffering', after observing his reactions when he dosed himself with an extract of Peruvian bark. The bark, which contains quinine, a substance Hahnemann knew was an antidote to malaria, produced symptoms similar to those of the disease it was meant to cure. This led to Hahnemann's 'law of similars' and to the principle underlying homeopathy that like cures like (*similia similibus curantur*), which he proposed in 1796.

Homeopathy was introduced into Britain, despite vehement opposition, by Dr Frederick Foster Harvey Quin. The fact that Quin was personal physician to Queen Victoria's uncle Prince Leopold, combined with the near-hysterical outcry from the medical profession, helped to publicise his cause. Quin founded the first homeopathic hospital in London's Soho in 1850.

PUTTING ON THE PRESSURE

As well as inserting needles into the body at selected points, the ancient Chinese applied pressure at the same places to cure illness. In reflexology, another ancient Chinese therapy, pressure is applied to the feet, where zones on the soles are thought to correspond to different body regions and organs. In 1913 the zones were introduced to the West when the US consultant William H Fitzgerald devised a system of massage that he called zone therapy. His ideas were further developed in the 1930s by Eunice D Ingham, who concentrated attention on the feet.

Chiropractic and osteopathy
Restorative manipulation

Manipulating the body to cure diseases was practised by the Egyptians in the 17th century BC but was not revived until 1895 when a Canadian-born doctor, Daniel David Palmer, cured the deafness of Harvey Lilliard from Iowa by manipulating the vertebrae in his neck. Through this act the system of chiropractic was founded. It was named by a clergyman, one of Palmer's patients, from the Greek *kheir*, or 'hands', and *praktikos*, or 'practical'.

Osteopathy was devised by the American doctor Andrew Taylor Still during the American Civil War. Combining his knowledge of anatomy and his interest in engineering, Still investigated how physical problems might be caused by misalignments of the body, and worked out treatments by manipulating and realigning patients' spines and other parts of their skeletons.

Aromatherapy
Beneficial oils

Early in the 20th century the French chemist René Gattefosse, whose company produced concentrated or essential oils for cosmetics, burnt his hand. He plunged it into some lavender essence and it healed rapidly, leaving no scar. After using essential oils on casualties during the First World War, Gattefosse became convinced of their effectiveness and in 1928 coined the word 'aromatherapy' for his treatments.

In Britain the use of concentrated oils as antiseptics and medicines began in the 13th century. It later became widespread, but essential oils fell out of use when weaker versions were made in the 19th century. Marguerite Maury, the French biochemist and beautician, was the first to use the oils in therapeutic massage, in the 1950s.

Family planning
Being in control

The Kahun papyrus of ancient Egypt, dating to about 2000 BC, contains the earliest written contraceptive advice: it advocated inserting gum, a mixture of honey and sodium carbonate, or a paste of sour milk and crocodile dung into the vagina. Another Egyptian prescription involved moistening a lint tampon with honey and ground leaves of the acacia, a plant known to contain lactic acid, which is deadly to sperm.

Aristotle, the ancient Greek philosopher, wrote that women 'anoint[ed] that part of the womb on which the seed falls with oil of cedar, ointment of lead, or frankincense commingled with olive oil' to prevent conception. In modern times Marie Stopes, the pioneer of family planning, found that olive oil did indeed have some spermicidal effects.

In the 2nd century AD the Roman physician Galen listed plants, including juniper, and plant extracts such as the bitter spice asafetida, that women could swallow as oral contraceptives. But most such ancient potions were ineffective, unpleasant and often dangerous.

Reliable oral contraception became possible only after the menstrual cycle was fully understood. Aëtio of Amida, a 6th-century Byzantine scholar, had realised that conception could be prevented if intercourse took place only on 'safe' days of the month. In 1927 the Austrian physiologist Ludwig Haberlandt extolled the virtues of 'hormonal sterilisation based on biologic principles'. But public opinion remained opposed to the idea, despite the work of Marie Stopes in Britain and Margaret Sanger in the USA.

Sanger, who coined the phrase 'birth control', had opened the first birth control advice centre in Brooklyn in 1916.

The cap, or diaphragm, was originally recorded by a German physician, F A Wilde, in 1823. It was later improved by rubber technology. Casanova, the 'Latin lover' of the 18th century, is said to have used halved lemons as caps to protect his conquests from pregnancy.

CONTRACEPTIVE CHOICES

In the 1950s scientists reopened their studies of birth control. In 1956 the American researchers Gregory Pincus and John Rock published the results of their clinical trials of the contraceptive pill. When it became available in the early 1960s the Pill was believed to fulfil Pincus's prerequisites as 'harmless, entirely reliable, simple, practical, universally applicable and aesthetically satisfactory to both husband and wife.' The yam, used for centuries as an oral contraceptive by Mexican women, played an important part in the development of the Pill. It contains a chemical used to make the female hormone progesterone.

The first barrier contraceptives used by men were the sheaths applied by the Romans as much to guard against disease as pregnancy. These were probably made from animal gut as were later condoms, named after the 17th-century Dr Condom, who is thought to have been a physician to Charles II. Such condoms, of which the best came from sheep intestines, were used until 1829 when the American inventor Charles Goodyear made rubber ones practicable. The Durex brand was launched in 1932 by the London Rubber Company.

Childbirth

Going into labour

When a woman went into labour her mother or another older woman would attend her as a midwife. Propriety excluded men. Women in ancient Egypt gave birth in a crouching or kneeling position or in chairs, and ancient Greek women sat on stools. The Romans had birthing chairs, commonly used in Britain until the 18th century.

From the 7th century BC until at least the 17th century AD women wore eagle stone amulets to protect themselves during pregnancy and childbirth. The naturally occurring *aetites*, 'pregnant' stones with sand or a pebble inside them, were used for the same purpose by the ancient Assyrians.

By the late 18th century qualified 'male midwives' had begun to practise. The Chamberlens, a French Huguenot family of male midwives whose forebears had fled to England in 1569, were using forceps in the 17th century. The success of the Chamberlens was built on such instruments, and women were even blindfolded to stop them seeing these powerful, mysterious tools.

MEDICAL INTERVENTION

When women began giving birth in hospital, from the 18th century, many died from puerperal sepsis. Charles White, the British obstetrician, appreciated the need for cleanliness to prevent the disease in the 1730s. But it took more than a century before any significant advance took place. Ignaz Philipp Semmelweis, who began work in Vienna in 1844, noticed that women whose hospital deliveries were attended by students were at greatest risk. The problem, he realised, was that students had often come straight from dissecting cadavers of women who had died of puerperal fever.

From 1847 Semmelweis insisted that those examining women during and after labour should wash their hands in a solution of chlorinated lime. Two years later the death rate had plummeted to about one in a hundred. Semmelweis was derided, but before the century had drawn to a close his theories had been accepted by the British surgeon Joseph Lister, pioneer of antisepsis.

Different forms of anaesthesia were used to ease the pain of labour from the mid 19th century; Queen Victoria inhaled chloroform while giving birth to Prince Leopold, her eighth child, in 1853. As she recorded later in her journal: 'Dr Snow gave that blessed Chloroform and the effect was soothing, quieting and delightful beyond measure.'

Opthalmic surgery

Operating on the eyes

In attempts to restore failing sight the Babylonians probably performed the first eye operations in about 1800 BC, by pushing lenses clouded by cataracts further into the eye. The Babylonians may have passed their knowledge to the Indians and to the Romans. Specialist Roman eye doctors regularly removed ingrowing eyelashes. The most skilled among them also used syringe-like needles to treat cataracts. In his medical writings of the 1st century AD Celsus stressed the great care required: 'the assistant from behind holds the head so that the patient cannot move; for vision can be destroyed permanently by slight movement …'

The itinerant oculists of the Middle Ages also acted as eye physicians, but only after the German physician Georg Bartisch wrote about eye diseases in the 16th century did ophthalmology start to become a speciality. The first formal courses in ophthalmology were held in 1803 at the University of Göttingen, although these were predated by the first surgery for strabismus or squint, carried out in 1738. Ophthalmology became more precise after 1851, when Hermann von Helmholtz, a German physiologist, invented the ophthalmoscope to create a magnified image of the retina at the back of the eye, where clues to disease can be found.

EARLY INSIGHTS

- Concave lenses to aid far vision appeared in a portrait of Pope Leo X painted by Raphael in 1517.
- The Egyptians made artificial clay eyes. Glass was used in Italy in 1578.
- The Inuit put slitted pieces of wood in front of their eyes to protect them from glare. Tinted sunglass lenses were available in the 1700s.
- Italians called early spectacle lenses *lenticchie*, or 'lentils', because of their resemblance to lentil seeds.
- In 1921 James Biggs of Bristol painted his stick white to draw attention to his blindness.

Improving vision

The first spectacles

In about 1280 an illiterate Italian glazier was inspired to place a pair of convex glass discs in front of the eyes to make near objects look clearer. Single lenses were already used in magnifying glasses by this time, but these eyeglasses were revolutionary. In the earliest written reference to them, an Italian named Sadro di Popozo commented in 1289: 'I am so debilitated by age that without the glasses known as spectacles, I would no longer be able to read or write.'

Early wearers used cords or chains to hold spectacles in place, otherwise, they rested on or gripped the nose. The earliest spectacles found in Britain, made in about 1500, were among the first with pads to help to secure the frames on the bridge of the nose. Glasses with rigid side pieces were originally made in 1727 by Edward Scarlett, a London optician. Bifocals were invented by Benjamin Franklin, the American statesman. In about 1785 he had the lenses from different pairs of spectacles sawn in half then cemented together to avoid having to swap glasses for close work and seeing at a distance.

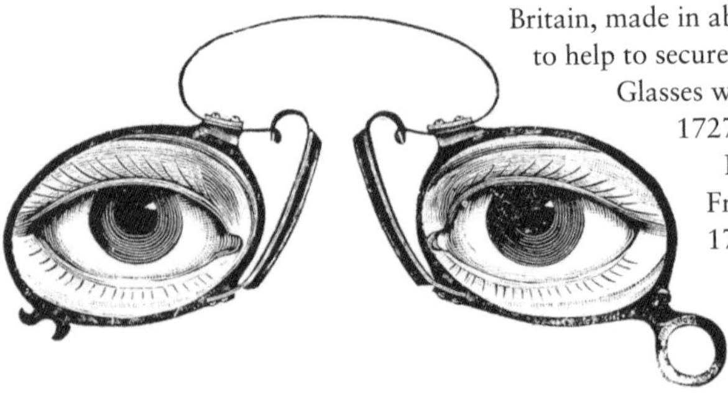

In 1827 the English physicist John Herschel proposed the idea for contact lenses, but they did not become a reality until 1887, when the Swiss physiologist Adolf Eugen Fick had a pair made by a glass-blower to correct astigmatism. But it was not until 1933 that a practical version was introduced, by the German ophthalmologist Josef Dallos.

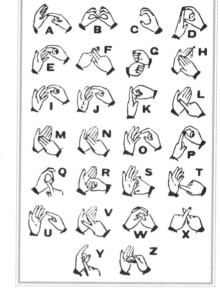

Help for the blind
Guides and alphabets

One day during the First World War a Dr Gorlitz, director of a sanatorium, was walking with a partially paralysed officer in the garden. Their stroll was interrupted when Gorlitz was called away, but sensing that the officer was in distress the director's dog, an Alsatian named Excelsior, ran indoors to fetch a walking stick. By the time Gorlitz returned, the dog was carefully guiding the soldier home.

Gorlitz was convinced that dogs could be trained to care for the blind. An exploratory programme was set up in 1916 to assist German soldiers who had been blinded in battle. The first guide-dogs were trained by the German Association for Serving Dogs and the Austrian War Dog Institute.

While he was still a teenager the Frenchman Louis Braille, who had been blind since the age of three, devised a system using patterns of up to six raised dots to represent letters of the alphabet. Braille's idea stemmed from the 12-dot battlefield communication system invented for night-time use by Captain Charles Barbier, a French army officer, in 1819 – the year Braille became a student at the Paris school for blind children.

Hearing aids
Acoustic assistance

To boost their hearing, ancient peoples funnelled sounds into their ears with animal horns. The simple ear trumpet was still commonplace in Victorian times – often skilfully disguised. Men would hide one below a beard, while women sported 'acoustic' fans.

In 1901 the Acousticon was patented by Miller Reese Hutchinson of New York. It was a large box containing batteries and electric circuitry, including valves to amplify sound. An attached device similar to a telephone receiver was held to the ear. Queen Alexandra, who had been partially deaf since childhood, used an Acousticon during King Edward VII's coronation in 1902. Smaller, portable hearing aids were made practicable by transistors from the 1950s.

The dental profession
Treating the teeth

Early dentistry must have been a particularly painful experience. The removal of diseased teeth was common from Egyptian times, when any teeth that were firmly fixed in the jaw were usually knocked out with a stone. Looser ones were pulled free with the fingers.

For centuries dentistry was practised by anyone who could acquire the necessary skills, and in medieval times pulling teeth was the job of barber surgeons, blacksmiths and horse doctors – or anyone, really, who felt capable of tackling the task.

The Egyptian sage Hesy-Re carried out the first recorded dental work in the 3rd millennium BC, which probably included both rudimentary fillings and extractions. 'Operators for the teeth' extracted the rotten molars of 17th-century Londoners and made dentures for the wealthy, thereby beginning modern dental practice. Britain's first specialist was Peter de la Roche, dentist to the royal household. His patients included the wife of the diarist Samuel Pepys. On March 11, 1661, Pepys recorded what was probably scaling: 'At night home and find my wife come home; and among other things, she hath got her teeth now done by La Roche; and are endeed now pretty handsome …'

The Frenchman Pierre Fauchard, whose definitive reference work, *The Surgeon Dentist*, was published in 1728, effectively made dentistry a science and advocated dental training. By 1760 British practitioners had begun to copy the French and call themselves dentists. Eighty years later in the USA, the world's first dental school was founded in Baltimore, Maryland.

Tools of the trade
Pull, drill, fill and clean

The Romans pulled teeth with strong iron forceps and used finer bronze ones for tidying up any fragments. They also employed dental mirrors and probes. Modern versions of these devices date only from the 19th century.

A tooth from a 4,500-year-old skull discovered in Denmark bears the marks of a slim flint drill that had once relieved an abscess, but drills were rarely used until the early 1700s, when the French dental pioneer Pierre Fauchard employed a jeweller's bow drill. Strength and speed were added to the armoury when John Greenwood, dentist to George Washington, adapted a treadle-powered spinning wheel to turn a dental drill in 1790. The hand drill was invented in 1850 by Charles Merry, also in the USA.

The teeth of Egyptian mummies reveal resin fillings, and resin, wax and coral were used to plug cavities up to the Middle Ages. Gold, which is easy to work, resistant to wear and nontoxic, was used in the Middle Ages. Fillings were regarded as symbols of wealth and they remained the preserve of the rich until the 18th century, when tin and lead were pressed into filed or drilled cavities as substitutes for gold.

Mercury was mixed with copper or silver to make the first amalgam in the early 1800s. Modern silver amalgam fillings, also containing mercury, were the 19th-century invention of Thomas W Evans, who was born in the USA but became dentist to the French emperor Napoleon III.

Before the early 18th century, when comfortable wooden armchairs were first provided, patients sat on the floor. The dentist worked while gripping the patient's head between his knees. Dental chairs with padded headrests, devised by the American Josiah Flagg, were first made during the 1790s. Attached to James Snell's reclining chair of 1832 were a mirror and a spirit lamp to shine a light into the mouth.

Bristled toothbrushes were made in China in the late 11th century, but in Britain they did not go on sale until 1649. Roman toothpowders contained emery to remove stains, but they also abraded the tooth surfaces, leaving blackened stumps. Chalk-based toothpastes were devised in the 19th century. The first tube of toothpaste, called Dr Zierner's Alexandra Dentifrices, went on sale in London in 1891.

THIS WON'T HURT A BIT

To protect themselves against the agonies of toothache, the ancient Egyptians rubbed burnt mouse dung and burnt angleworms into their troublesome teeth. One ancient Roman cure was to apply olive oil in which earthworms had been boiled. Until safe anaesthetics arrived, alcohol was used to numb the pain of dental surgery. In 1846 William Thomas Morton of Boston, USA, extracted a tooth from a patient who had been rendered unconscious with ether. Nitrous oxide, also known as laughing gas because of its effects, was preferred from about the turn of the century.

False teeth
A new set of gnashers

The first known false teeth were made by Neferites, who practised in ancient Egypt around 2600 BC. From 700 BC the Etruscans made partial dentures by mounting extracted teeth, or teeth carved from bone, ivory or ox tooth, on gold bridges supported on each side by healthy teeth. These false teeth were unmatched until the 19th century.

A pair of dentures dug up from a field in Switzerland are the oldest full set yet discovered; with teeth carved from bone and attached with gut to a hinged side piece, they were probably made in the 15th century. Early dentures with wooden plates, especially upper sets, were notoriously hard to keep in place, and in some 18th-century designs they were secured with springs that pressed against the gums. It was vulcanised rubber, invented by Charles Goodyear in 1839 and the plates moulded from it from the 1850s onwards, that eventually provided a snug fit.

The first false teeth with the looks and strength of the real thing were the porcelain dentures made by the French apothecary Alexis Duchâteau in about 1770. Many other materials had previously been tried, including ivory which, though durable, made the breath smell. George Washington put his ivory set in port overnight to help to improve both the taste and the odour.

Index

Acknowledgments

The Brooking Collection: 11 BR, 20 TL, 24 BL
Mary Evans Picture Library: front cover TR, BL, BC, BR, 9, 58, 76-77, 84 L, 92, 97 T, 104 TR, 108, 115, 121 T, 136, 141, 151, 152-153, 170, 180, 189, 198, 199, 200, 201, 202-203 R, 205, 221 B, 229, 230 BL, 236-237, 242 TL, 245, 247 TR, 247 BR, 248
Mary Evans Picture Library/Edwin Wallace: 208 TL
Mary Evans Picture Library/© Illustrated London News: front cover TC, back cover BR, 14, 28, 68, 142 TL, 171
Mary Evans Picture Library/Interfoto: back cover T
Mary Evans Picture Library/©Thomas Cook Archive/Illustrated London News: 202 L

All other illustrations come from Art Explosion images; Dover Publications, Inc., Mineola, New York; Harrods Ltd Victorian Shopping Catalogue, 1895; Heal's Catalogue 1853-1934

Project Editors Caroline Boucher, Caroline Smith
Art Editor Simon Webb
Designer Keith Davis
Cover Design Sailesh Patel
Additional editorial assistance Bina Taylor,
Ollie Grieve
Introduction Ruth Binney
Proofreader Ron Pankhurst
Indexer Marie Lorimer

Reader's Digest General Books
Editorial Director Julian Browne
Art Director Anne-Marie Bulat
Managing Editor Nina Hathway
Trade Books Editor Penny Craig
Picture Resource Manager Sarah Stewart-Richardson
Pre-press Account Manager Dean Russell
Product Production Manager Claudette Bramble
Production Controller Jan Bucil

Origination FMG
Printed and bound in China

The Curious History of Everyday Things
Published in 2012 in the United Kingdom by
Vivat Direct Limited, (t/a Reader's Digest),
157 Edgware Road, London W2 2HR

The Curious History of Everyday Things is owned
and under licence from The Reader's Digest
Association, Inc.

The Curious History of Everyday Things has
previously been published under the title
Extraordinary Origins of Everyday Things (2009).

We are committed both to the quality of our products
and the service we provide to our customers. We value
your comments, so please do contact us on 0871 351
1000 or visit our website at www.readersdigest.co.uk

If you have any comments or suggestions about the
content of our books, email us at
gbeditorial@readersdigest.co.uk

ISBN 978 1 78020 124 5
Book Code 400-595 UP0000-1